BIOTECHNOLOGY

W0245866

The ILO's World Employment Programme (WEP) aims to assist and encourage member States to adopt and implement active policies and projects designed to promote full, productive and freely chosen employment and to reduce poverty. Through its action-oriented research, technical advisory services, national projects and the work of its four regional employment teams in Africa, Asia and Latin America, the WEP pays special attention to the longer-term development problems of rural areas where the vast majority of poor and underemployed people still live, and to the rapidly growing urban informal sector.

At the same time, in response to the economic crises and the growth in open unemployment of the 1980s, the WEP has entered into an ongoing dialogue with the social partners and other international agencies on the social dimensions of adjustment, and is devoting a major part of its policy analysis and advice to achieving greater equity in structural adjustment programmes. Employment and poverty monitoring, direct employment creation and income generation for vulnerable groups, linkages between macro-economic and micro-economic interventions, technological change and labour market problems and policies are among the areas covered.

Through these overall activities, the ILO has been able to help national decision-makers to reshape their policies and plans with the aim of eradicating mass poverty and promoting productive employment.

This publication is the outcome of a WEP project.

Biotechnology

A Hope or a Threat?

Edited by

Iftikhar Ahmed
Development Economist, International Labour Office, Geneva

Foreword by

Michael Lipton
Institute of Development Studies, University of Sussex

Palgrave Macmillan

ISBN 978-1-349-12867-9   ISBN 978-1-349-12865-5 (eBook)
DOI 10.1007/978-1-349-12865-5

© International Labour Organisation 1992
Softcover reprint of the hardcover 1st edition 1992

All rights reserved. For information, write:
Scholarly and Reference Division,
St. Martin's Press, Inc., 175 Fifth Avenue,
New York, N.Y. 10010

First published in the United States of America in 1992

ISBN 978-0-312-07154-7

Library of Congress Cataloging-in-Publication Data
Biotechnology : a hope or a threat? / edited by Iftikhar Ahmed :
foreword by Michael Lipton.
    p. cm.
Includes bibliographical references (p.   ) and index.
ISBN 978-0-312-07154-7
1. Biotechnology—Social aspects.   I. Iftikhar Ahmed, 1944– .
TP248.2.B523   1992
303.48'3—dc20   91–47618
                                                      CIP

To my wife Selina and daughters Alvina and Shadia

Contents

PART III   MILK AND FEED

PART IV   THE EXTERNAL THREAT

Foreword
Michael Lipton

This important book aims to analyse, predict and thereby improve the impact of biotechnology (BT) on poor people in developing countries. Modern BT comprises the alteration of the genetic structure of an organism, not by selection – as accomplished by nature for millions of years, by farmers for thousands, and by scientists for a few score – but by direct genetic intervention, viz. tissue culture or gene transfer. So far, BT has concentrated on human health, industrial processing, cattle, soil bacteria and dicots, in that order. Its impact on poor people via crop production in developing countries, however, depends quite heavily on the *type* of BT – anther culture, to speed up purity (homozygosity) in parents for F1 hybrids; meristem culture, to produce disease-free clones of roots and tubers; and so forth. The linkages between type of crop, BT and poverty reduction are explored in detail in my book with Richard Longhurst, *New Seeds and Poor People* (pp. 364–83).

Although the main impact of BT has not been on tropical food staples, the yields of cereal crops in many parts of the Third World had been doubled – often tripled – between 1964 and 1988, largely before BT, by another dramatic set of innovations: the so-called Green Revolution (GR) in traditional plant breeding. BT's main potential contribution to the production of food staples is probably as an adjunct to, and an accelerator for, traditional plant breeding. This will continue to be the mainstay of food crop improvement for several reasons. Many crucial crop traits – for example, drought – are affected by several genes in complex ways; direct manipulation of such genes is not yet feasible. Even when it is, effective selection will depend on the response of BT planting materials to various environments, that is, to the tests of traditional plant breeding, first in the research station, then in farmers' fields. Indeed, there is a real danger that BT will increasingly divert scientists and funds away from the more 'traditional', but central, plant breeding and testing that lie at the heart of crop improvement.

Nevertheless, this book demonstrates that our analysis of BT's effects on the poor needs to learn from the experience of GR. Both BT and GR can enrich many poor people through higher and more stable employment,

nutrition and small-farm income. Both can harm the poor if they are
directed, by policies or by markets, mainly towards labour-displacing
activities, or towards improving the competitiveness only of farms or
regions that tend to operate less labour-intensively – in other words, of
the more prosperous farms or regions. Yet many critics of the GR in the
1970s, and of BT today, wrongly concentrate their fire on the technology
itself, not upon its misuse and misdirection.

The main danger with BT – even more than with the GR – is that poor
people's regions and products may be neglected, rather than that they are
directly harmed. The GR began with wheat and rice in well-watered areas;
in several of these, notably the Indian Punjab, the GR has helped to make
poverty much rarer. The GR has also lately provided major benefits for
some crops of very poor people – sorghum, finger millet, even cassava – in
some rainfed areas. It is the vast areas where food staples are untouched
by GR, above all in most of Africa, that poverty has worsened. Yet
the current targets of BT – while including cheaper or new industrial
processes, often displacing labour-intensive Third World products such
as sugar or (potentially) beverages – largely neglect monocots, including
cereals. Much crop production by poor people, especially in remote areas,
is thus at risk of being by-passed, rather than being actively harmed or
distorted, by BT as by the GR.

Yet the potential for better results exists. In Mexico (Chapter 4) BT, by
rapidly developing and spreading disease-free coffee and coconut plants,
could probably help poorer producers in the very regions (and crops) left
out by the GR. Due care is needed, of course, to avoid over-rapid extraction
of scarce nutrients, especially from impoverished soils. BT, like the GR,
needs to concentrate its search upon planting materials with high and safe
'conversion efficiency' of nutrients into dry matter, and 'partitioning efficiency' of dry matter into economic products – not 'extractive efficiency'
at soil mining. However, this book shows that such projects from BT – for
very poor farmers in Kenya, Malawi, and the Philippines – are possible,
and justify the editor's warning to avoid 'negative generalisations which
could deprive' the poor of high potential 'benefits of BT'.

In developed countries, the heavy concentration of early BT – for example via
bovine growth hormone – upon highly organised and science-linked animal production tends to favour big farms. In developing countries, however,
Table 11.3 (based on Kenyan data for tea and potatoes) suggests that
older BT – just like the GR – has strong potential to reinforce small farmers'
advantages. Although bigger farmers may adopt first, once adoption is
well advanced both BT and the GR re-confirm the 'inverse relationship'
in which small farmers have higher output-per-hectare than large farmers.

Both BT and the GR, however, show larger farmers scoring better for labour productivity (i.e. worse for employment in the short term.)

A growing majority of the world's poor depend mainly on wage employment, not income from farmland. Unlike the GR, some BT displaces fertilisers, somewhat reducing employment in the Mexican case. Elsewhere, however, this is outweighed by extra employment in irrigation, harvesting, and local industrial linkages. Employment of weeding labour, usually very poor women, is sometimes increased by BT. However, such employees are severely threatened by systemic herbicides linked, via BT, to resistant and high-yielding cotton or maize hybrids.

Perhaps the main benefit from the GR for the poor was more ample, reliable, inexpensive flows – and stocks – of food staples. The effect of BT on consumption and nutrition is little explored. In developed countries, BT may well improve the health effects of fats, for example, by reducing erucic acid in rapeseed. In the developing world, the little BT research done on food staples tends to move with market demand, not always with consequences that favour the reduction of poverty. For example, BT concentrates on yellow maize to feed poultry rather than white maize to feed poor Filipinos (Chapter 9).

In analyzing the potential of BT to improve human nutrition, some misperceptions must be avoided. The great majority of the world's undernourished people require more calories; a small minority require more protein. Both majority and minority are undernourished because they are very poor. The very poor can hardly ever afford much costly animal-based food. Therefore more – or, even, somewhat cheaper – milk, beef or poultry is of little direct use to the poor. In Nigeria, however, single-cell protein as animal feed (if cheapened by BT) may indirectly help the poor by allowing land to be switched from soybeans for animals to cheaper calorie sources for poor people (Chapter 8)

Such indirect effects, however, are likely to be of secondary importance. Until BT can be applied to tropical food staples, poor people's nutritional gains from BT depend mainly on its capacity to bring them higher employment income. This capacity, except in tropical food staples, is very uncertain, being indirect and, especially, dependent on international economic equations and relationships. For example, BT's emphasis on developed-country cattle productivity can well improve the balance of payments of Third World importers of dairy products from, and/or exporters of feed to, First World Cattle; but big employment effects are not very likely. Moreover BT, if it continues to concentrate on livestock production, could shift land even in the Third World from cereals to cattle – reducing employment and raising the average cost of calories.

As for BT's apparently large potential impact in raising Third World outputs of cash crops, the position is less clear. Some of these crops, like cotton in the many developing countries that are net importers, are largely consumed as well as produced by the poor; here, BT clearly helps the poor. This is also the case for cash crops consumed largely in the developed world – if they are consumed in conditions of price-elastic demand, and produced labour-intensively, as with some vegetable oils. However, BT that raises output of price-inelastic Third World crops consumed mainly in developed countries – tea, coffee, perhaps palm oil – might benefit only a few early adopters, while harming the Third World as a whole, especially poor employees, as the glut leads to a price crash. The main beneficiaries would be consumers in the First World; Third World income from these commodities, especially that of poor employees, would fall.

Despite the great local promise of BT, its effect in significantly raising incomes for the Third World's poor people *as a whole* probably depends mainly on steering it more towards the food-crop staples that they grow, work on, and eat. This is made harder by the fact that, for several reasons, BT research – unlike GR research – has so far concentrated on activities where the benefits from progress can be 'internalised' to the researchers, or their employers, sufficiently to produce a commercial return. This has, for example, concentrated crop BT on organisms, varieties or populations not readily reproduced and retained by the original recipients, for example, on outbreeding crops rather than on inbreeders such as wheat and rice; and on F1 hybrid seeds which the farmer must buy fresh each year, rather on F6 inbreeders or on composite populations of outbreeders, where farmers can often retain their seeds. Thus the drive towards profit – in many walks of life a major source of poverty reduction, as well as of growth – has led in the case of BT research to the privatisation, and hence by definition to the undersupply, of the non-price-excludable public good that is comprised by most forms of research-based improvements in tropical food staples; and to the distortion of such improvements *towards* price-excludable, and hence profitably privatised, forms.

Evidence from the Philippines (Chapter 9) confirms that the major shift of agricultural research towards private, market-seeking suppliers in the BT era – unless complemented by appropriate public action – militates against effective, rapid access to BT innovations by small, poor and especially subsistance farmers, above all by those who eat much of what they grow. More generally, private research is not normally motivated by social returns. The GR – despite some ill-informed criticism – did help many poor farmers, workers and consumers; initially directed towards their main food staples, it was researched and developed by public-sector – often

international public-sector – scientists who, although sometimes exces-
sively motivated by inappropriate peer-group criteria, were to a significant
extent protected from pressures to produce results mainly for the farmers
(and consumers) with most market power. BT, potentially more powerful
and more durable even than the GR, is *not* initially directed mainly
towards food staples, and *is* mainly researched by Western private-sector
researchers for large firms required to meet profit criteria.

The history of private research into hybrid maize in the United States,
among many other events, shows that *competitive* private profit seeking – especially alongside an overview, and research competition, by a
public sector itself kept accountable – can help poor farmers and consumers by cutting costs, even in the provision of a public good like
much agricultural research. The risk with BT, however, is the gradual
strengthening of monopolistic private suppliers of particular BT products
that cannot be reliably reproduced, or profitably avoided, by farmers
themselves. Such suppliers are located overwhelmingly in the rich world.
Public overview is limited to negative, sometimes damaging, environmental restraints. Public-sector agricultural research itself – which could
provide competition (and fill gaps) – increasingly lacks resources, information, and control instruments either to steer private researchers towards
social priorities or to meet them publicly. Meanwhile, private BT and other
activities increasingly attract away the public sector's scientists.

These problems are most serious in the Third World. All but a few of its
governments (Brazil, China, India, Mexico, Pakistan and one or two more)
lack resources, scale, capacity or political demand for public-sector BT on
a significant scale. International agricultural research, which was crucial
for the GR varieties of tropical food staples, faces at most 'level funding'
and is not obviously best placed to divert it substantially from traditional
plant breeding (which will be much more important in food production for
at least ten years) to either basic or applied BT.

Potentially and over the next few decades, BT can transform the livelihoods of the world's rural poor. The realisation of this hope depends,
not on penalising private BT, but on providing massive public-sector (or
at least non-profit-seeking, as in the case of the Rockefeller Foundation's
important initiative on rice BT) competitors, overviews, and, above all,
complementing counterparts that focus on tropical food staples. Such a
focus should concentrate upon improved planting materials that make
better use of water, thereby 'substituting employment for environment'
in fragile lands, whether semi-arid or ill-drained.

However, even if developing countries (big ones and/or consortia
of smaller ones) and international research agencies get these extra

resources today, a decade or more will elapse before poor Third World food employees, farmers and consumers reap major benefit. Meanwhile, the local and largely cash-crop initiatives documented in this book are welcome and necessary. However, they should be focused on activities and areas that minimise the danger that all, or more than all, the benefits accrue to consumers of tropical products in the rich world.

Acknowledgements

This volume reflects the efforts of a number of individuals who have contributed in various ways at different stages of the work. Ajit Bhalla, Gerd Junne, Wouter van Ginneken and A. K. Ghose provided many useful comments.

I am very thankful to Susan Saidi and Micheline Batailley who helped with the tedious work of placing the entire manuscript on the word processor.

IFTIKHAR AHMED

List of Contributors

Iftikhar Ahmed is a Development Economist with the ILO World Employment Programme and has worked on agricultural and rural development. Previously, he was a Post-Doctoral Associate at the Iowa State University of Science and Technology, a Visiting Fellow at the Institute of Development Studies, University of Sussex, and Associate Professor of Economics, Dhaka University, Bangladesh. He is author of *Technological Change and Agrarian Structure: A study of Bangladesh*, co-editor of *Farm Equipment Innovations in Eastern and Central Southern Africa*, editor of *Technology and Rural Women: Conceptual and Empirical Issues* and co-editor (with Vernon W. Ruttan) of *Generation and Diffusion of Agricultural Innovations: The Role of Institutional Factors*.

Paolo Bifanni is Professor of Environment and Development at the Universidad Autónoma de Madrid and has undertaken major assignments for UNCTAD, WIPO, InterAmerican Development Bank, ILO, FAO and other international bodies on the subject of biotechnology, environment and technology, and international trade development. He served in a senior position with UNEP, Nairobi. He has numerous significant publications in these fields.

Chinyamata Chipeta is a Professor of Economics at Chancellor College, University of Malawi and Head of the Department of Economics. During 1978–81 he was an Assistant to an Executive Director at the International Monetary Fund. He has worked as a consultant to several international organisations including the World Bank, the Southern African Development Co-ordination Conference (SADCC), the United Nations Economic Commission for Africa (ECA) and UNESCO. He is the author of *Indigenous Economics* and *Economics of Indigenous Labour* as well as numerous journal articles and contributions to books.

Amarella Eastmond is a researcher in the Social Science Unit of the Autonomous University of Yucatán (Mexico) and is interested in the socioeconomic impact of new technologies on the peasant producers of the region. She is currently carrying out fieldwork on the orange growers in the south of Yucatán for her Ph.D. thesis. She has written various papers on general socioeconomic issues of plant biotechnology in Mexico.

Dr. Saturnina C. Halos is a geneticist at the University of the Philippines serving in various capacities, mainly as a senior researcher at the Natural

Sciences Research Institute, Diliman Campus; seconded researcher at the National Institutes of Biotechnology and Applied Microbiology (BIOTECH), Los Baños Campus; visiting professor of genetics, Institute of Biological Sciences, Los Baños Campus and lecturer in microbiology and cell biology, Institute of Biology, Diliman Campus. She has worked with several government agencies in planning research programmes in support of development projects in agriculture and forestry since 1978. She has twelve years experience in biotechnology research, mainly plant tissue culture and genetic improvement of industrial micro-organisms.

Harold H. Lee is a Professor of Biology and Director of the Master of Liberal Studies programme at the University of Toledo, Ohio, United States. His scientific expertise is in large-scale animal cell culture technology with two US patents on cell culture instrumentation and with research interest in animal fertilisation and embryogenesis. He has lectured extensively on biotechnology, especially on embryo engineering and developmental biology. Since 1986, he has been a member of the Network of the Advanced Technology Assessment System of the United Nations Centre for Science and Technology for Development. He has served as a consultant for governmental and industrial concerns for training of workforce, research laboratories and pilot plants for cell tissue culture technology. He has published in the area of cell biology, biotechnology and university curriculum modernisation.

Boniface F. Makau is currently Principal Economist, Ministry of Planning and National Development, Kenya. He was previously Science Secretary (in charge of agricultural sciences) at the National Council for Science and Technology. His main topics of interest include resource productivity and mobilisation. He has conducted some studies on the returns to research.

Mary Wezi Mhango is a lecturer in Home Economics at Chancellor College, University of Malawi. She obtained an MA degree in Home Economics Education in 1984 at Mount Saint Vincent University, Canada and B.Sc. in Home Economics in 1974 at the University of Missouri, United States. Her research interest and experience is in income generation activities for women and small business for rural communities.

Leopold P. Mureithi is an economist with interest in the areas of human resources, technology, industrialisation and rural development. He is currently an Associate Professor of Economics at the University of Nairobi, Kenya. His publications include Crisis and Recovery in African Agriculture, Development Options for Africa, Food, Population and Rural

Development in Kenya and Employment, Technology and Industrialisation in Kenya.

Gilbert U. Okereke is a Senior Lecturer and Project Leader of Agricultural Biotechnology Research Laboratory of Anambra State University of Technology, Enugu, Nigeria. Currently he is National Secretary of the Biotechnology Society of Nigeria and one of the editors of the *Nigerian Biotechnology Journal*, and Dean, Faculty of Agriculture at Anambra State University of Technology. He was the Head, Department of Food Science and Technology of the University from 1981 to 1983. He has participated in many national and international conferences in the area of biotechnology in food production.

Gerardo Otero is an Assistant Professor of Latin American Studies at the Simon Fraser University, Burnaby, British Columbia, Canada, prior to which he was a Visiting Senior Lecturer at the University of Wisconsin, Madison. He was previously Professor-Researcher at the Instituto de Estudios Económicos y Regionales of the University of Guadalajara in Mexico and a research consultant at the Instituto Tecnológico y de Estudios Superiores de Occidente. He was a post-doctoral research fellow at the Center for U.S.-Mexican Studies at the University of California, San Diego. He has contributed papers and chapters to professional journals and edited volumes on the subjects of agrarian structure and biotechnology. He is currently completing a book on political class formation in rural Mexico.

Manuel L. Robert is Head of the Division of Plant Biology of the Centre for Scientific Research of Yucatán where he is in charge of basic and applied research programmes on plant biotechnology. He has been involved in developing plant biotechnology methodologies for private industry. Apart from specialised papers on plant biotechnology, he has co-edited a book on plant tissue culture in Mexico and contributed chapters to books on different aspects of biotechnology.

Guido Ruivenkamp is Director of the Research Department on Biotechnology and Agriculture of 'Land-en Tuinbouw Bond' (LTB) in Haarlem (the Netherlands). Previously he has worked at the TNO Centre for Technology and Policy Studies, where he was jointly responsible for the research on socio-economic effects of biotechnological applications in food production and agriculture. His Ph.D. Research at the University of Amsterdam was on 'The introduction of biotechnology in the agroindustrial application chain of production; changing over to a new form of labour organisation'. Currently, he is coordinating research on the biotechnnological developments for the agricultural sector in the Netherlands and in developing countries. He has numerous publications in this field.

Frederick E. Tank is Chairman of the Department of Economics and Associate Director of Economic Education at the University of Toledo. Dr. Tank received his doctorate from Wayne State University in 1972 with a specialisation in urban and regional economics and public finance. He has written numerous scholarly articles on a wide variety of topics (his latest publication which appeared in the *Journal of Macroeconomics*, 1988, deals with the utilisation of vector autoregressive techniques to analyse the determinants of investment within a macroeconomic model). He is developing a modelling procedure to estimate the impact of agricultural biotechnology by product category.

List of Abbreviations

ADC	Agricultural Development Corporation
ACIAR	Australian Centre for International Agricultural Research
APB	Advanced Plant Biotechnology
APC	Average propensity to consume
ASTI	Advanced Science and Technology Institute, Philippines
BANRURAL	Banco Nacional de Crédito Rural (National Bank of Rural Credit)
bGH	Bovine Growth Hormone (the technical term is bovine somatotropin, BST)
BIOTECH	National Institute of Biotechnology and Applied Microbiology, University of the Philippines
BR	Bio-Revolution
BST	Bovine somatotropin
CIP	International Potato Centre (Centro Internacional de la Papa)
CONACyT	Consejo Nacional de Ciencia y Tecnología (National Council of Science and Technology – Mexican Federal Government)
COSNET	Consejo del Sistema Nacional de Educación Tecnológica (Council of the National System of Technological Education)
DCs	Developing Countries
DNA	Deoxyribonucleic Acid
ELISA	Enzyme-linked Immunosorbent Assay
EAP	Economically Active Population
FAO	Food and Agriculture Organization of the United Nations
FDA	United States Food and Drug Administration
GAO	General Accounting Office of the United States Congress
GATT	General Agreement on Trade and Tariffs
GEPLACEA	Grupo de Paises Latinoamericanos y del Caribe Exportadores de Azúcar (Group of Latin American and Caribbean Sugar Exporting Countries)

GR	Green Revolution
HFCS	High Fructose Corn Syrup
HYV	High Yielding Variety
IFPRI	International Food Policy Research Institute, Washington D.C.
IRRI	International Rice Research Institute
ISNAR	International Service for National Agricultural Research, The Hague
KGGCU	Kenya Grain Growers Cooperation Union
KSh	Kenya shilling
KTDA	Kenya Tea Development Authority
LDC	Less Developed Country
LY	Leathal Yellowing
MFS	Mimiosa field selection
MLO	Mycoplasma-like Organism
MP	Micropropagation
MPS	Marginal propensity to save
MV	Modern variety
NPQS	National Plant Quarantine Station
NPRS	National Potato Research Station
NSO	National Statistical Office
PCARRD	Philippine Council for Agriculture and Resources Research and Development
PCI	Progeny Clone I
PH	Hydrogen ion concentration
PRONASE	Productora Nacional de Semillas (National Seed Producer – Mexican Federal Government)
PTC	Plant Tissue Culture
PVC	Polyvinyl Chloride
RAFI	Rural Advancement Fund International, Manitoba, Canada
R&D	Research and Development
rDNA	Recombinant Deoxyribonucleic Acid
RNA	Ribonucleic acid
SARH	Secretaria de Agricultura y Recursos Hidráulicos (Ministry of Agriculture and Water Resources – Mexican Federal Government)

SCP	Single Cell Protein
SFS	Swazi field selection
S&T	Science and technology
STA	Smallholder Tea Authority, Malawi
TC	Tissue Culture
TF	Tea Flavour
TRF	Tea Research Foundation, Kenya
UNAM	Universidad Nacional Autónoma de México (National Autonomous University of Mexico)
USDA	United States Department of Agriculture
US-OTA	United States Office of Technology Assessment
VP	Vegetative propagation

List of Tables

List of Figures

1 Introduction and Overview

Iftikhar Ahmed

BACKGROUND

Agriculture is the main source of income of the world's poor. The greatest numbers of poor people, including the very poorest, live in the rural areas. It is not surprising that their livelihoods are linked to the efficiency and productivity of agriculture, whether or not they derive their incomes directly from it (World Bank, 1990).

In general, the rural poor belong to wage labour or marginal farmer households. Their poverty results as much from low returns as from unemployment and underemployment. The numbers of the rural poor have increased from 767 million in 1970 to 850 million in 1985 (Singh and Tabatabai, 1990).

On a global basis, between the early 1960s and 1980s, world food crop production grew a half per cent faster than the growth in population (Mellor, 1988). Despite the above positive margin of growth in food production over population growth, the absolute number of undernourished people in Third World countries actually increased from 460 million in 1969–71 to 512 million in 1983–85 (Singh and Tabatabai, 1990). This increase in the number of hungry people cannot be attributed to the variability of food production during this period.[1]

Hunger results from the inability of poor countries, poor families and poor individuals to purchase sufficient quantities of food from available food supplies. In the above context, it is evident that biotechnology applications could make a contribution to poverty-alleviation if they could boost poor people's purchasing power by improving labour absorption in rural areas without sacrificing growth in agricultural output. In this respect biotechnology could also enhance food security, particularly in sub-Saharan Africa by its possible contribution to increasing food supply.

Furthermore, the agricultural labour force in the Third World is projected to increase at 0.8 per cent per annum during the years 1985 to 2000. This estimate conveys to policy makers the quantitative magnitude of additional jobs to be created in agriculture, especially if the scope for creating the required non-agricultural and urban employment opportunities is limited (Singer, 1991).

Apart from such important social considerations, there are many more powerful reasons why Third World countries need biotechnologies to pull them out of agricultural stagnation and propel them to higher growth rates, and to narrow or diminish the enormous North-South technological gap. *Firstly*, a yield plateau now appears to have been reached for major crops, particularly for rice[2] (Barker, 1989). Yields from rice varieties introduced since the Green Revolution have remained unchanged around the potential of the high yielding IR-8 rice variety released in 1984 (Lipton and Longhurst, 1989). Therefore, future yield gains critically depend on what biotechnology applications have to offer. *Secondly*, the pace of technological change in agriculture has been slow. For example, technological change in agriculture doubled corn yields *every 30 years*. In contrast, microelectronic technologies helped double productivity *every 1.1 years*. Even the jet aircraft technology doubled productivity *every 13 years* (Molnar and Kinnucan, 1989). Thirdly, biotechnology is desperately needed by the Third World countries to reduce the huge North-South technological gap in the agricultural sector. This is clear from the fact that labour productivity in the industrial sector of developed countries is only four to six times higher than that of the developing countries. In contrast the labour productivity in the agricultural sector of developed countries is 15 to 22 times higher than those of the developing countries (ILO, 1991). *Fourthly*, plant agriculture alone accounts for 61 per cent of the potential biotechnology market, conservatively estimated to be worth US$50 billion a year. The next in importance is chemicals, with 20 per cent of the total potential market, then come human medicine (10 per cent), food ingredients (4 per cent), animal husbandry (2 per cent) and aquaculture (1 per cent) (UNCSTD, 1984). *Fifthly*, purely from an analytical point of view, the agricultural sector offers a previous technological experience (the Green Revolution) which provides a comparative backdrop for evaluating the potentials of the new biotechnologies. *Sixthly*, focus on a single sector limits the varying definitions and concepts of biotechnology which can become necessary when analysing issues across sectors, and so reduces the complexities of economic forecasting.

WHAT DOES THE SOCIAL SCIENCE LITERATURE OFFER?

The scientific literature on biotechnology is indeed very robust but it was not until the past several years that social scientists have made any effort to introduce a socioeconomic dimension to the international debate on this subject. A set of twelve books (Table 1.1) now available can be divided into

TABLE 1.1   *Profiles of the newly emerging social science literature on biotechnology 1987–1990 (major books)*[1]

Author(s)/editor(s)	Title	Publisher	Year of publication
	Global/cross-sectoral reviews		
OECD	*Biotechnology: Economic and wider impacts*	OECD, Paris	1989
Edward Yoxen and Vittorio di Martino[2]	*Biotechnology in future society: Scenarios and options for Europe*	Dartmouth Aldershot	1989
Albert Sasson	*Biotechnologies and development*	UNESCO, Paris	1988
A. Sasson and V. Costarini	*Biotechnologies in persepctive: Socio-economic implications for developing countries*	UNESCO, Paris	1991
RIS	*Biotechnology revolution and the Third World: Challenges and policy options*	RIS, New Delhi	1988
Robert Walgate	*Miracle or menace? Biotechnology in the Third World*	Panos Publishing Ltd, London	1990

TABLE 1.1 (continued)

Author(s)/editor(s)	Title	Publisher	Year of publication
	Agriculture Sector Reviews		
Henk Hobbelink	*Biotechnology and the future of world agriculture*	Zed Books, London	1991
Joske Bunders	*Biotechnology for small-scale farmers in developing countries: An analysis of assessment procedures*	VU University Press, Amsterdam	1990
Gabrielle Persley	*Agricultural biotechnology: Opportunities for international development*	ISNAR, The Hague	1989
Joseph J. Molnara and Henry Kinnucan	*Biotechnology and the new agricultural revolution*	Westview Press, Boulder, Colorado	1989
John Farrington	*Agricultural biotechnology: Prospects for the Third World*	ODI, London	1989
David Goodman, Bernardo Sorj and John Wilkinson	*From farming to biotechnology: A theory of agroindustrial development*	Basil Blackwell, Oxford, UK	1987

1. For abbreviations see the bibliographical reference at the end of the volume.
2. Coverage of this book is narrowly confined to issues relating to Europe only.

two categories: (a) global treatment of the subject and (b) agriculture sector studies. Much of the work (which remains fragmented and scattered) treats partially of the socioeconomic issues.

The global reviews cover too much ground and too many issues across diverse sectors, resulting in generalisations which are not very helpful for specific policy making. For instance, Sasson's *Biotechnology and Development* deals with biotechnologies for agriculture, forestry, livestock husbandry and animal health, medicine and public health, production of pharmaceuticals, energy and environmental pollution. Even the agriculture sector studies are based on sketchy data and socioeconomic analysis that is inadequate and lacks penetration. Most of the socioeconomic literature is speculative and conjectural, often resulting in the blurring of issues. This volume in contrast, deals concretely with specific agricultural biotechnologies and, as a result, treats the socioeconomic issues more rigorously in much greater depth. The thrust of our discussion is on job creation and poverty alleviation, and in this context the central socioeconomic issues are carefully defined. The conclusions arrived at here, often based on the actual testing of hypotheses, are more meaningful not only for policy-making and for the formulation of future programmes but for the setting of priorities for biotechnology development and application. It is to be recognised that much of the criticism of the current biotechnology development relates more to the direction or orientation (on socioeconomic issues) of the R & D rather than on its pace. Since the social scientists remained inactive in the past, it is not surprising that biotechnology R & D is more science-led than guided by socioeconomic considerations.

METHODOLOGICAL AND THEORETICAL CONSIDERATIONS

The blurring of issues on account of the sweeping generalisations and lack of clarity and depth of analysis in the social science literature on biotechnology is a direct consequence of certain methodological complexities and shortcomings which impair empirical research. Therefore, the special methodological skills which integrate the biological sciences with the social sciences are briefly discussed below. This is followed by a discussion of the methodological problems of socioeconomic forecasting arising from some conceptual ambiguities.

Innovative methodologies

Empirical research for forecasting the socioeconomic and employment impact of the new biotechnologies is severely handicapped by serious data limitations: (a) reliable data are not available in sufficient quantities to reflect the production response to biotechnologies not yet released or disseminated (see Tables 1.2 and A.1); and (b) existing data on the characteristics of the breakthrough are simply not made available for policy analysis because the information is in the hands of the private sector which is reluctant to supply commercially sensitive statistics. This poses a major challenge to social scientists, who are usually fed *ex post* or farm management survey data on which to base their analysis.

Our analyses attempt to provide useful analytical insights to scientists, entrepreneurs and policy-makers since they have to make both production and research decisions regarding the new biotechnologies. The primary objective is to anticipate and mitigate or avoid the adverse effects and capitalise and enhance the positive benefits of the newly emerging biotechnologies. Data generated by geneticists and microbiologists, though very fragmented and scattered, are harnessed to the extent possible in this volume for *ex ante* socioeconomic analysis of the consequences of these new biotechnologies using innovative methodologies. Essentially, this requires very special skills in combining an in-depth knowledge of the technical aspects of the subject with a broad appreciation of the socioeconomic conditions prevailing in the Third World countries.

It is, therefore, not very surprising that *ex ante* assessment research has assumed special significance for the newly emerging biotechnologies. However, the limitations of *ex ante* research should be recognised: (a) the future multi-disciplinary consequences of biotechnologies are difficult to predict by the work of any one discipline (e.g. for handling the legal intricacies of proprietary rights over living entities); (b) the methodology tends to be more sophisticated (e.g. the need for use of general equilibrium approaches to study inter-industry repercussions); (c) the vested interests (e.g. the large multinational companies) may not be pleased with the results of *ex ante* assessments if the predictions are unfavourable to them, the valid socioeconomic findings being dismissed on flimsy grounds and their impact being unpredictable; and (d) reliable data are more difficult to generate at the micro level simply because producers can provide information on their current production or cost structures more accurately and quantitatively than they can for the situation arising five or more years in the future (Fishel, 1987).

In the light of the above discussions, this volume follows a mixed methodological approach. Analyses in Chapters 5, 6 and 8 (Chapter 7 partially) are based on hard data generated by country case studies (*ex post* survey data) which permit the empirical verification of numerous hypotheses discussed below. In the rest of the chapters the volume

TABLE 1.2 *Availability of new biotechnologies for selected Third World crops, 1989*

Crop	New diagnostics[1]	Rapid propagation systems[2]	Transformation systems[3]	Regeneration systems[4]	Time-frame for commercial applications (number of years)
Banana/Plantain	+	+	−	+	5–10
Cassava	+	+	−	−	5–10
Cocoa	+	−	−	−	>10
Coconut	+	−	−	−	>10
Coffee	+	+	−	+	5–10
Oilpalm	+	+	−	−	>10
Potato	+	+	+	+	0–5
Rapeseed	+	+	+	+	0–5
Rice	+	+	+	+	0–5
Wheat	+	+	−	−	>10

1. Availability of new diagnostics for pests or diseases based on the use of monoclonal antibodies or DNA probes.
2. Availability of rapid propagation systems to allow the multiplication of new varieties.
3. Availability of transformation systems to enable new genetic information to be inserted into single plant cells.
4. Availability of regeneration systems to enable single cells to be regenerated into whole plants, after transformation.

SOURCE: Persley, 1989, p. 23, Table 3.1.

relies heavily on deductive reasoning, matching specific biotechnology breakthroughs in industrialised countries to concrete socioeconomic needs. *Ex ante* assessment is obviously unavoidable when the world is on the threshold of biotechnology breakthroughs for the major Third World crops (Tables 1.2 and A.1). As a consequence, many of the conclusions may be conjectural or speculative.

Given the time-frame for many of the newly emerging advanced biotechnologies for crops of importance to the Third World (Table 1.2), much of the forecasting of the probable impact would naturally have to be based on deductive reasoning using simulation approaches.

Conceptual ambiguities

Another major difficulty faced in the assessment and projection of the socioeconomic impact of biotechnology are the varying definitions used for biotechnologies. Since this leads to ambiguities of interpretation the practical value of the findings on impact of biotechnologies to the policy-maker is greatly eroded. Moreover, the social scientist is often baffled by the near-indecipherability of the often terribly complicated jargon used by the geneticists. The essays in this volume, therefore, make a special effort to translate the complex scientific aspects of biotechnology into simple language: understanding of these concepts is often essential to a proper grasp of the issues and a full appreciation of the socioeconomic consequences of biotechnolgical innovation. (To assist the reader a special glossary of technical terms has been included in the beginning of the volume).

The gradient of biotechnologies is provided in Figure 1.1 in an ascending order of their scientific complexity and sophistication. While Chapter 3 analyses the alternative concepts of biotechnology, the unique feature of biotechnology definition used throughout this volume refers to a cluster of commercial techniques which employ living organisms to make or modify a product, including techniques for improving the characteristics of economically important plants and animals and for developing micro-organisms which act on the environment (UNCSTD, 1984). More specifically, Chapters 4, 5, 9 and 10 deal in varying degrees with micropropagation (tissue culture) techniques often labelled as 'second generation' biotechnologies. Chapters 8 and 9 deal with micro-organisms used both in pre-engineered and engineered contexts. Chapter 7 deals exclusively with a genetically engineered product (BST). Chapter 11 contains socioeconomic analysis based on an additional survey of transgenic (i.e., containing a foreign gene) plants and microbes.

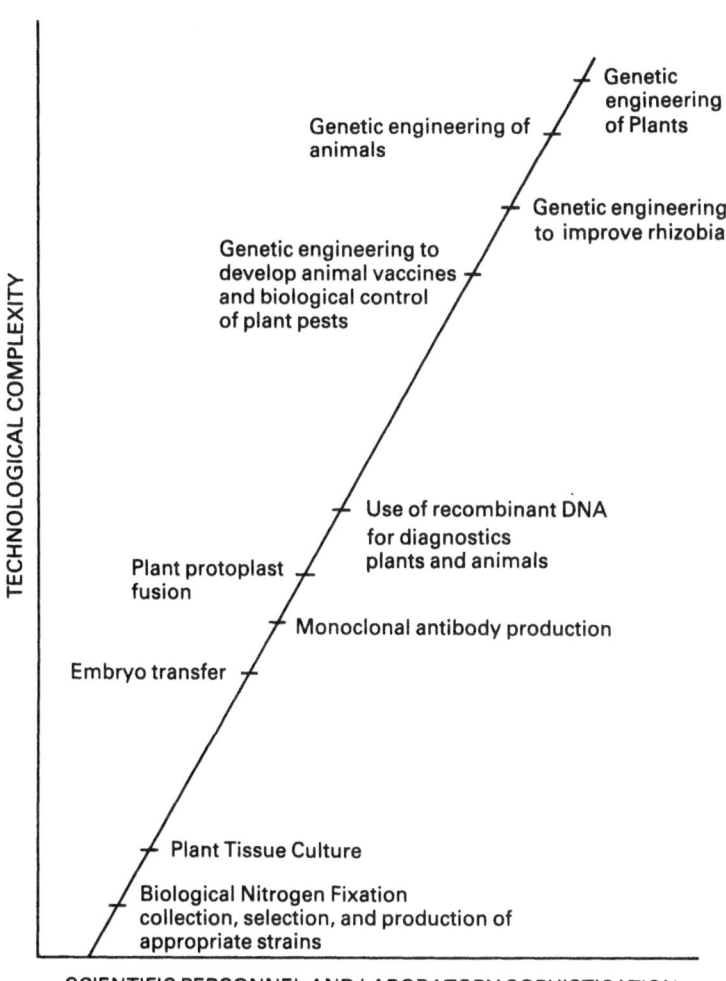

Figure 1.1   Gradient of Biotechnology

Source: Adapted from Jones (1987).

PURPOSE

Our overall approach has been to seek and analyse the positive elements arising from the current and future biotechnology developments. The policy maker's attention is also drawn to biotechnologies which may contain socially harmful consequences. Our overall purpose, therefore, is to survey the biotechnologies and to alert the policy-makers and practitioners of their socioeconomic consequences in good time. For this purpose, we analyse some policy options which could mitigate the adverse impact, and enhance the beneficial impact, before biotechnologies are developed and released for commercial application.

The contributions to this volume advance some hypotheses covering the socioeconomic impact of biotechnologies. In order to provide credible advice our contributors attempt the empirical verification of a set of these hypotheses which are built essentially around major socioeconomic issues affecting the various countries that we analyse. Wherever feasible, the important hypotheses corresponding to individual issues and themes, especially those that have been tested, are indicated. In this way, it is hoped that this volume will generate some analytically useful *ex ante* information to the policy makers and entrepreneurs to guide them in making their immediate policy- and production-decisions. We also provide signals to the geneticists and microbiologists on the types of data urgently needed by economists to conduct more meaningful socioeconimic analysis, in particular in areas identified for future socioeconomic research.

ISSUES AND HYPOTHESES

We argue in this volume that the integration of biological and physical sciences with the social sciences is a prerequisite for an understanding of the potentials and application of biotechnologies in solving the problems of the Third World poor.

The first set of hypotheses has been formulated to seek answers to some important questions relating to farm size. For instance, will the transgenic plants (i.e., containing a foreign gene) and genetically-engineered microbes be potentially more scale neutral (i.e. equally applicable to large as well as small farmers) at the farm level than the Green Revolution and mechanical innovations? Will the cost-reducing or output-enhancing potential of biotechnologies be more beneficial to small farmers (for their profits or their survival)?[3] At the national cross-sectoral or macro level do resource-saving agricultural biotechnologies tend to depress GDP and reduce aggregate employment? Do the newly emerging biotechnologies have the potential for reducing previously uncontrollable production variances in the agricultural sector?[4] The answer to this latter question is of particular concern to the risk averse, resource-poor small farm sector. Will the structural adjustment measures with a movement towards the free market (price liberalisation, withdrawal of agricultural input subsidies and greater reliance on the private sector for the marketing and distribution of inputs) induce biotechnology innovations? More specifically, it is important to determine (where labour is cheaper relative to agrichemicals) whether the factor allocation decisions of peasants induce them to adopt biotechnologies so that they can replace agrichemicals by labour, without sacrificing (or even increasing) output or net revenue.

The next set of hypotheses relates to the employment implications of biotechnology, covering important issues like changes in the labour intensity in agriculture, structure and stability of rural employment, and the impact on the rural labour market, particularly on wage labourers and women workers. The magnitude and skill composition of jobs created, and the non-farm employment generated through the forging of backward linkages to the laboratories, and forward linkages to the marketing and crop processing sectors as a result of biotechnology applications in agriculture, are also analysed.

We also seek answers to a set of questions on biotechnology development policy which could contribute to poverty alleviation. Can biotechnologies be specifically designed and deliberately released to alleviate rural poverty? In this connection, our discussions attempt to demonstrate, perhaps a bit dramatically, how biotechnologies can be skilfully deployed to launch a planned assault on solving specific socioeconomic and technical problems in economically depressed and ecologically harsh rural areas.

Essentially, the central question under consideration is: do biotechnology developments need to be more objective-led rather than science-led if these are to be socially beneficial? Indeed, the essays gathered here transmit definite signals and a very important message to scientists and plant breeders. It is argued that the genes that are transferred by genetic engineering should ultimately be seen as a social question if this technology is to truly provide answers to the problems of mass poverty and widespread and growing unemployment.

Another major issue concerns the new international division of labour resulting from the disruptions in global trading patterns. Do biotechnology developments in the industrialised countries affect both North-South as well as South-South trade? How permanent are these changes? What categories of the Third World population are likely to be affected by these changes? Is a major restructuring of the economies of affected countries essential? What other countries are threatened with loss of markets for traditional exports?

The volume warns against the current popular negative perceptions of biotechnology which have led to adverse generalisations which could deprive the Third World countries of biotechnology's positive benefits. Are all external biotechnology developments (in the industrialised countries) socially harmful to developing countries? How can Third World countries' access to socially beneficial, but patented, biotechnologies be improved? Do Third World countries have the institutional and technological capability to adapt the new biotechnology to suit the local socioeconomic environment? Are existing Third World technological and institutional capabilities in the simpler 'second generation' biotechnologies rightly targeted?

Because of its exclusive concern with the socioeconomic dimensions, we do not deal with the legal aspects of the debate on intellectual property rights, although it does address the question of Third World farmers' access to these technologies. Another area that has to remain outside the scope of our discussion concerns the environmental consequences of the deliberate release of transgenic plants and microbes, even though there is an awareness of the controversy over this issue.

DESIGN OF THE VOLUME

In the light of the background, methodology, objectives and hypotheses outlined above, the volume brings together conceptual and empirical analyses undertaken by economists and sociologists with the aid of, and often in partnership with, microbiologists. With somewhat varying emphasis the volume adopts the dual approach of *ex ante* deductive reasoning and of tailoring their arguments to the type and availability of hard data (i.e. *ex post* survey data).

The book is divided into four distinct parts. Part I develops the issues raised in this chapter. In addition, Chapter 2 formulates and applies a simulation model to measure and trace in an input-output framework the effects of resource-saving agricultural biotechnologies on aggregate output and employment. The analyses in Parts II and III and Chapter 11 also lead to different degrees of conceptualisation on the basis of *a priori* reasoning and analysis of empirical evidence. Part II is exclusively concerned with plant biotechnologies. This part clearly distinguishes between the two methodological approaches of deductive reasoning and use of hard data. Chapter 3 applies the former approach in which socioeconomic and technical problems faced by the small farm community in a non-Green Revolution area are clearly identified. A set of *feasible* biotechnologies are analysed for their contribution to the solution of these specific problems and for their contribution to income generation, job creation and poverty alleviation. In contrast Chapters 4 and 5 adopt the conventional approach of economic analysis since the authors use *ex post* or farm management survey data to fuel their analytical engines.

Part III contains case studies which deal mainly with animal-based products. Chapter 7 is special in two ways: (a) empirically it examines the differential impact of a given biotechnology in the two contrasting socioeconomic settings of developed and developing countries; (b) methodologically, in the case of the former (United States), where the biotechnologies in question (Bovine Somatotropin and High Fructose Corn Syrup) are being commercially applied, hard data was available for analysis. In the case of the latter (Mexico), the analysis had to rely on *ex ante* assessment of the biotechnologies through deductive reasoning. Chapter 8 is based largely on *ex post* survey data specially collected with the purpose of analysing empirically the actual socioeconomic impact of SCP in the Nigerian context.

Part IV analyses the effects of biotechnology developments on international trade flows which in turn drastically affect the international division of labour. Since these changes lead to massive displacement of labour in the plantation sector and the loss of livelihood of small farm holdings this is considered as a major external threat by virtually all the developing countries. This part also has the advantage of drawing upon actual cross-country production and trade statistics.

Chapter 11 not only sums up the major issues raised in this volume, but also reinforces the conclusions and findings with additional empirical evidence. It places the facts before the policy makers, donors, the private industry and the academic community for positive action since each one has an important role to play on biotechnology decision-making. Major research gaps are identified, especially in the area of rural labour markets. In addition, a revolutionary research agenda is proposed in this concluding chapter which calls for aggressive policy intervention. It also emphasises the need to mobilise political will as well as financial and scientific resources, to fully exploit the potential of biotechnologies for poverty alleviation.

NOTES

1. Indeed during the 1980s food production in the developing market economies has not become more unstable over time (ILO, 1990).
2. Yield plateaus arise because crops stubbornly refuse to increase their yields no matter how much extra fertiliser is applied to them, and also because the work of the plant breeders is unable to cope with the spread of new diseases.

3. Theoretically, the volume views this question in terms of cost functions where farmers are assumed to select (on the cost curve) the least-cost combination of inputs to produce a given level of output. Indeed, the purpose of technological change is to reduce the use of inputs at any relative price so that those scarce resources can be re-allocated by the farmer to other uses.

4. This hypothesis is proposed mainly because the insect and disease-resistant biotechnologies counter some of the stress and pathological losses associated with disease and insect infestations. It is primarily this characteristic of the new biotechnologies which could contribute to the reduction in the previously uncontrollable fluctuations in production.

Part I
Conceptual Approaches

2 A Conceptual Framework for Biotechnology Assessment

Harold H. Lee and Frederick E. Tank

INTRODUCTION

This chapter examines the economic impact of biotechnology, the social consequences of these economic changes, and some of the policy prescriptions that might lessen the undesirable consequences which could accompany the commercialisation of biotechnology and the application of bio-products. Our analysis will focus on the redistribution of surplus farm workers, resulting from the introduction of yield-enhancing agri-biotechnology, into expanded or newly created industries. We argue that social and economic stability can be maintained by expansion of existing or through the establishment of new industrial infrastructure. The workers who are employed in new industries will provide a vital link between the primary producers and consumers. These industries also forge an essential link between the development of biotechnology, the use of the innovations by the farmers and the ultimate benefits that will be derived by the members of the individual households.

GREEN REVOLUTION VS. BIO-REVOLUTION: A BRIEF ACCOUNT

Using molecular biological techniques to create 'new species' or to modify existing biological forms is the basis for what is called the Bio-Revolution (BR). Well before the advent of the BR, thanks to the Green Revolution (GR), there was a marked increase in the production of food in many of the less developed countries (LDCs). These increases stemmed primarily from the efforts of plant breeders working at two international research centres who developed high-yield varieties of wheat and rice (for a review, see Swaminathan, 1982 and Sau, 1988). Along with the use of herbicides, fertilisers, pesticides and modern farming equipment, these new plants

17

concurrently increased global food production as well as the productivity of some LDC farmers. The most important component of the BR will be similar to the major result of the GR, i.e. increased food production. Both revolutions encompass changes in various biological systems which result in increased production from old agricultural inputs or supplementary production from new inputs. In both the GR and BR, biological and other technological innovations are used to enhance both the quality and quantity of outputs.

The GR and the BR are, however, significantly different in several important aspects. The major dissimilarities occur in the requisite occupational skills, the scientific expertise needed for Research and Development (R & D) and even the application of the bio-products. Biotechnology incorporates the scientific principles of microbiology, biochemistry, genetics, and biochemical and chemical engineering. The fundamental technologies which are behind BR are therefore much more complex, requiring the knowledge of many disciplines, all of which have sizeable educational requirements for minimum competence. Additional dissimilarities include the sources of raw materials and capital equipment, the ease of technology transfer and the resultant spin-off or 'satellite' industries. The support/supply industries for those firms actually engaged in the production of biotechnical products will also be quite different from the kind of firms necessary to support the end-products of the GR. The GR was much more open; in other words, there was open exchange of know-how among nations in both public and private sectors to foster its implementation. The primary constraints on output (yields) in the GR were, for the most part, the physical inputs themselves, that is capital equipment, available land and irrigation systems, along with the required financing to purchase these inputs (Sau, 1988). On the one hand, the BR is constrained by the highly technological and specialised training (Kanawaty, 1985) required not only to develop new products, but also to devise the means to mass produce the bio-products in a cost-effective fashion. On the other hand, the GR was involved much more in conventional agricultural methods such as traditional selective breeding, cultivation, weeding, and so on, which are, for the most part, labour-intensive. In many LDCs including South Asia, the division of labour for these activities is culturally determined (Ahmed, 1987).

In some countries (particularly, but not exclusively, the United States) the governmental regulatory process is a significant impediment to the commercial development of biotechnology-related agricultural products. Extremely cautious, unresponsive and confusing regulatory mechanisms are frequently an impediment to the field-testing of bio-products, slowing and perhaps reducing the economic potential of BR. Obstacles of this nature were insignificant in the GR.

The GR had its greatest impact at the local or regional levels (Tangley, 1987), but these rather localised economic gains did not promote a rapid acceleration of overall economic development. The multiplier effect of these localised economic gains was, for the most part, quite limited in many LDCs. For example, the removal of the food constraint in India did not spearhead a surge in new industries, not even those that are food-related like the refrigeration and canning industries. As a result, very few individuals in the LDCs seem to have benefited from the GR *per se* (Sau, 1988). The living standards in India, Pakistan and the Philippines have been improved by the GR to the extent that more individuals have reached self-reliance at subsistence levels. On the other hand, Thailand and Indonesia seem to have benefited from the GR in a more dramatic fashion. Especially Indonesia was able to take advantage of its petro-resources, further enriching gains made possible by the GR. Like several other technologies, the BR can generate national and global consequences previously impossible because countries are engrossed in international trade in agricultural commodities at levels far above those of the GR (King, 1987). In addition, the science-based BR will be able to build on the advances brought about by the GR, which should have compounding consequences for agricultural activity (Swaminathan, 1982).

THE NATURE OF THE BIO-REVOLUTION

The interdisciplinary scientific basis of biotechnology

Biotechnology permeates many scientific specialities and is not exclusively the domain of biological sciences. To genetically engineer an organism, be it a microbe, plant or animal, requires knowledge and techniques from many scientific disciplines: biology, chemistry, biochemistry, physics, computer sciences, electronics, agriculture and animal husbandry, to name but a few. The expertise of scientists and technicians, together with traditional agricultural labour components, is required to take the inventions of biotechnology out of the laboratory and into the commercial markets. The continuing involvement of scientists is essential not only during the R & D period, but also in the bio-production processes and other phases which bring the outgrowths of the initial biotechnological research into the market place. A wide range of laboratory equipment, reagents and computer hardware and software are the productive inputs used in biotechnical research. These are developed and maintained by scientists and technicians from many disciplines. Therefore, the BR is not controlled exclusively by one or two areas of specific scientific endeavours, but relies on the research developments emanating from a co-operative and integrated effort of many individuals from many scientific areas.

The privatisation of biotechnology

Due to the rapid progress of molecular and cellular biology, there has been a dramatic broadening in the scope of potential applications and a quickening in the pace at which these products are brought through the initial research and development stages. With the emergence of these modern biotechnological methods, which lend themselves to commercialisation, the BR was expected to occur rapidly, with new and improved products, quickly outpacing the innovation rate of all other branches of science and technology in the industrialised nations (*Chemical Week*, 1986).

While many of the early inventions of biotechnology have emerged from universities and research institutes, there is a trend towards more and more private R&D. In the United States and most of the industrialised nations, firms engaged in R&D finance their own operations through public equity and private investment, including large amounts of venture capital. This implies that the technical know-how in the BR is increasingly proprietary. Biotechnology has moved away from being a purely research-oriented field in which the primary incentives are strictly academic. Biological knowledge and materials are no longer shared freely among all interested scientists. Instead, the incentives are economic: for example, patenting for exclusive rights to use and license. Industry is rushing to commercialise biological specimens and processes developed through research collaboration with universities.

In 1965, private sector R&D accounted for about 55 per cent of the total public and private-sector research in support of the United States food system. Farmers would collect and replant their own seeds each season, eliminating much of the profit incentive for private firms to become involved in plant breeding for the purpose of developing new improved varieties of hybrid seeds to sell to farmers (Tangley, 1986). Moreover, state and federal governments in the United States were actively engaged in the development and distribution of new crop varieties. There has been a major turn of events with the arrival of biotechnology. Since the late 1960s and early 1970s, private R&D has grown much more rapidly than public-sector agricultural research with much of the research focused on product development.

Recent trends indicate that the pace of privatisation of biotechnology is likely to quicken. Institutional innovations, more effective regulation of patenting and future marketing, are providing additional incentives for private-sector investments in science-based biological technologies.

Environmental aspects of the BR

In the future, LDCs are not likely to engage in large-scale biotechnology research and development programmes with the intention of introducing the 'bio-innovations' into their respective economies. It is more likely that they will serve as markets for agricultural products researched, developed and produced in the more industrialised countries. Further complicating the picture, the great potential of recombinant DNA (rDNA) technology in agriculture, even in the industrialised countries, is far behind the advances made in the pharmacological area. The reasons for this are partly scientific, partly environmental and partly economic. To elaborate, a very large portion of the world's food crops are monocot grains like corn and wheat; and the *genetic engineering of these plants* is only at an early stage. Scientists have turned to genetic engineering not of the plants themselves but of the micro-organisms upon which plants depend, for example, the nitrogen-fixation bacteria in the soil. This has resulted in serious environmental concerns and the possible creation of costly protective procedures which will delay the field-testing of these microbes in several industrialised countries. The United States and European environmental lobbies such as the 'Gen-ethic Network' and the 'Green Party' vigorously oppose much biotechnical R&D and commercialisation of the results of biotechnology and have opposed the release of genetically-engineered organisms into the environment. The success of these opposition movements has forced biotechnology firms to build new industrial complexes and R&D facilities in places where the political 'climate' is more favourable or the environmental regulations are less restrictive.

While the genetic engineering of plants is less controversial than the creation of animals with the genes from non-parent animals (the transgenic animals), it is none the less very significant (Raines, 1988). World-wide environmental groups agree that, if anything, the biotechnology industry is under-regulated and that regulatory agencies are irresponsible in promoting a far too rapid advance in biotechnology applications. They believe that the current regulations are an uneven patchwork that is redundant in some areas and leaves large gaps in others. In the United States, for example, Jeremy Rifkin and his Foundation on Economic Trends has initiated legal action against government agencies to force environmental reviews (impact assessments) of all government biotechnology research. In Europe biotechnology firms are finding that, because of environmental concerns, production processes using rDNA micro-organisms are not receiving government approval and may be forced to move overseas. Firms in both Denmark and Germany have had problems assuring government officials that biotechnology waste from their production facilities will not be contaminated with recombinant organisms (*Biotech News*, 1987). Environmental issues on the release of rDNA organisms are not just confined to plant materials. The Wistar Institute of Philadelphia has conducted field trials of an rDNA rabies vaccine in Belgium (Goldstein, 1990), the same vaccine that caused an uproar in Argentina in 1986 when the Government was apparently not informed about the test that was taking place (*Science*, 28 October 1988).

One can summarise the environmental questions that are of concern. Can microbes engineered for specific plants or animals find their way via natural environmental interactions into other food crops? Or, can they cause a general deterioration of the ecosystem by some unanticipated biological relationship? The ecosystem is fragile and can even be weakened by seemingly minute alterations. Successful introduction into one ecosystem does not ensure that others, perhaps even in different countries, will not suffer adversely. While the consequences of environmental damage of this sort are difficult to predict, clearly the poor would not be able to escape the economic aftermath. Accordingly, LDCs should judiciously exercise particular caution in adopting the fruits of biotechnology since they may risk damaging valuable natural resources. Indeed, the multinational companies are seeking out Third World countries as convenient environments for field testing their transgenic plants (for the research directions of multinationals see Table A.1).

The environmental elements of the introduction of new technologies in LDCs need to be critically examined. Studies on ecological impact usually take a back seat to the analysis of the economic income-generating and productivity-enhancing effects. In developing countries, studies of the ecological and environmental impact of biotechnologically-produced organisms should not be made secondary to the income-generating aspect of their utilisation. Environmental assessments, despite their costly nature, are surely as important as economic assessments to make certain that short-term economic gains are not offset by long-term damage to the economic base and the precious natural resources. The consequences of ignoring these environmental aspects could evoke adverse effects on socioeconomic activities. To re-emphasise this point, one might engineer a microbe to enhance corn yields, but this same microbe may be harmful to woody plants. If that microbe spreads to the forests, the paper industry could be adversely affected. The so-called 'killer bees', even though they were produced by conventional genetic breeding techniques, provide another excellent parable warning us not to treat the environmental consequences of any new organisms lightly.

Application of biotechnology

One of the most compelling historical facts is that there are large differences in the ability of countries to generate technical innovations which improve that country's economic well-being. In addition, there has been significant variability in the willingness and ease with which societies have adopted and utilised technological innovations developed elsewhere (Rosenberg, 1982). The expectation is that the changes made possible through biotechnology will be able to cause actual alterations in equipment, products and organisation of the production. The diffusion of scientific knowledge surrounding advances in agri-biotechnology are particularly likely to exhibit a long lead over the eventual applications for several different reasons. There are many studies which suggest that profitable new technologies are adopted rapidly by farmers in both developed countries and in LDCs (Feder *et al.*, 1985). However, before this new technology can be utilised, it must be made available from the supply industries that are accessible. While the research activity in agricultural biotechnology is intense, product introductions at the commercial level will be slow in coming. In fact, there are not many genetically-engineered organisms being marketed to farmers, and the promised productivity enhancement, such as increased crop yields and pathogen and pest resistance, remain in laboratories (Vaughan, 1988). This is also evident from Table A.1. Although field tests of various genetically-engineered bio-products for use in the agricultural sector have been reported in the media, production at any significant scale, even for potential use by small privately-owned family farms, has yet to occur. This lack of commercial availability stems from numerous factors which will impede the rapid infusion of bio-products into LDCs. Some of the more important factors are discussed below.

Locational variations

Each crop occupies a special ecological niche. Whatever the crop – corn, wheat, rice, soybeans, alfalfa, etc. – there are usually special environmental forces that influence the growth of that crop. What may be a problem in Angola may not be a problem in Wisconsin. Locational factors include complex, interactive ecological systems prevailing in one region but not shared by others. Genetically-engineered drought-resistant crops may increase yields but, at the same time, may also increase the population of predators. This, in turn, may initiate a series of injuries to other resources which are difficult to reverse. The successful introduction of a technology therefore does not necessarily carry a connotation of economic success.

Market forces

Most of the R&D efforts of biotechnology firms in the industrialised countries are oriented to products which can be sold in more familiar domestic markets or the markets where people can afford to buy them. Since the profit margins in domestic markets tend to be more predictable than those in foreign countries, bio-products engineered specifically for a developing country are considered a riskier undertaking. Given that research in biotechnology is already a risky undertaking even in industrialised countries, further risk taking in the form of biotechnology geared for developing countries is currently very unattractive from a commercial viewpoint. Generally, the greater the uncertainty about market response and the ability of the organisation to absorb and utilise the requisite changes, the less likely it is that the technological innovation will be a success (Kline and Rosenberg, 1986).

The commercial make-up of any biotechnical innovation takes on increased importance as agricultural R&D moves out of the public domain and into the private sector. With the fruits of biotechnology being viewed as marketable commodities by business enterprises in developed countries, the transfer of this technology to the developing countries could take many different channels. Joint ventures between firms in LDCs and foreign biotechnology companies would allow participation and some control by the developing country, and, at the same time, satisfy any regulatory measures established by that government. The technology transfer might be accomplished through licensing agreements between multinational biotechnology enterprises and unaffiliated companies in the LDCs. The use of biotechnology licensing and other contractual agreements between firms in industrialised countries and LDCs tends to decrease the economic and technological dependence of these LDCs on the developed countries.

Research progress

Advances in animal science, particularly transgenic technology, will turn animals into 'chemical factories'. Genetic engineering has already transformed bacterial cells into 'mini-factories' producing useful pharmaceuticals, like insulin and BST. Moreover, scientists have introduced human genes into sheep in such a way that the protein products of the genes can be harvested from the animals' milk without inconveniencing the sheep. In addition to using the gene transfer technology to produce these proteins which are hard to obtain, researchers also employ transgenic technology to improve the characteristics of the animals themselves.

Reduced effective patent life and patent infringements

Reduced patent life resulting from longer periods of research and development, as well as lengthy government approval procedures, may decrease commercial viability. The average time required to develop a new biotechnical product far exceeds the time necessary to obtain a patent, lessening the effective life of the patent. This then decreases the profitability and thus the incentive to innovate. In general, a further disincentive to develop bio-products for the LDCs centres on the problem of patent protection. The difficulties in protecting patented products of biotechnology are much greater outside the country originating the patent. Developers of biotechnology products have attempted to find foreign licencees to market their innovations, but difficulties in winning approval of foreign patent claims make the licence agreements of uncertain value.

Public awareness

Companies engaged in biotechnical product development are becoming increasingly concerned with the public's perception of biotechnology, particularly the field-testing of genetically-engineered organisms. Negative publicity and heavy media coverage of public opposition not only tarnish the image of this new technology, but also impede its progress. Therefore, apart from assessing the scientific and economic impact of a new bio-product, many firms may have to ensure that the public is adequately informed about testing and safety procedures. Seeking approval from government or quasi-government concerns at locations where testing will be done is an essential step in product development. Educating the public through public-relations techniques can lead to a more positive image for biotechnology – another facet of the Bio-Revolution (*Science*, October 1988). For example, BioTechnica Agriculture, Inc., BioTechnica International, Inc., and Crop Genetics International, to name a few, conducted field tests without controversy due to their considerable efforts in informing the public. Public confidence in biotechnology can be increased if companies convey information to the general public about the precautions that are undertaken to ensure that potentially dangerous microbes, for example, are not released into the environment.

Technology transfer

Transfers of biotechnology to LDCs in the present era of the BR have been very limited. The restricted nature of technology transfer from the industrialised countries to the LDCs is somewhat surprising, since many of the gene sources originate from LDCs. In some ways, biotechnology itself resembles a product. A biological technique can be sold or the end-product of the bio-technology R&D, such as genetically-engineered plants or animals, may be marketed. In the latter case, the biotechnology is embodied in these genetically-engineered agricultural products or the capital equipment itself. Alternatively, the processes themselves (disembodied technology) can be transferred to foreign firms for remuneration. The transfer of biotechnical processes can be viewed as a disembodied transmission of specialised production and management know-how.

SOCIOECONOMIC CONSEQUENCES OF THE BIO-REVOLUTION

Appraisal of biotechnology infusion

One would expect that the introduction of advanced biotechnology into developing countries would be 'poor-friendly', i.e. it would raise the national income, the overall standard of living and particularly the economic welfare of the farmer. Successful infusion and implementation of biotechnology has the potential to alter the traditional economic structure. Agriculture would become part of a vast business network, the success of which will certainly have a positive effect on the economy. The most immediate effect illustrated in Figure 2.1, shows that higher yields per hectare would result in higher production levels. This could give rise to an agricultural surplus that could be exported to other regions or perhaps even another country. If regional output continues to be greater than regional demand, the excess harvest will need to be distributed to other regions by having strategically-located distribution networks – a new and necessary infrastructure if distant markets are to be reached.

With the creation of these distribution networks to handle the agricultural surplus, new avenues of employment will occur. New markets will be developed. The traditional direct sale from producers to consumers will diminish with changes in the structure of the food delivery system. These changes will therefore be 'poor-friendly' rather than 'anti-poor'. Because of the need for semi-skilled workers, such as truck drivers, mechanics and business clerks, it will be necessary to train and educate people for these

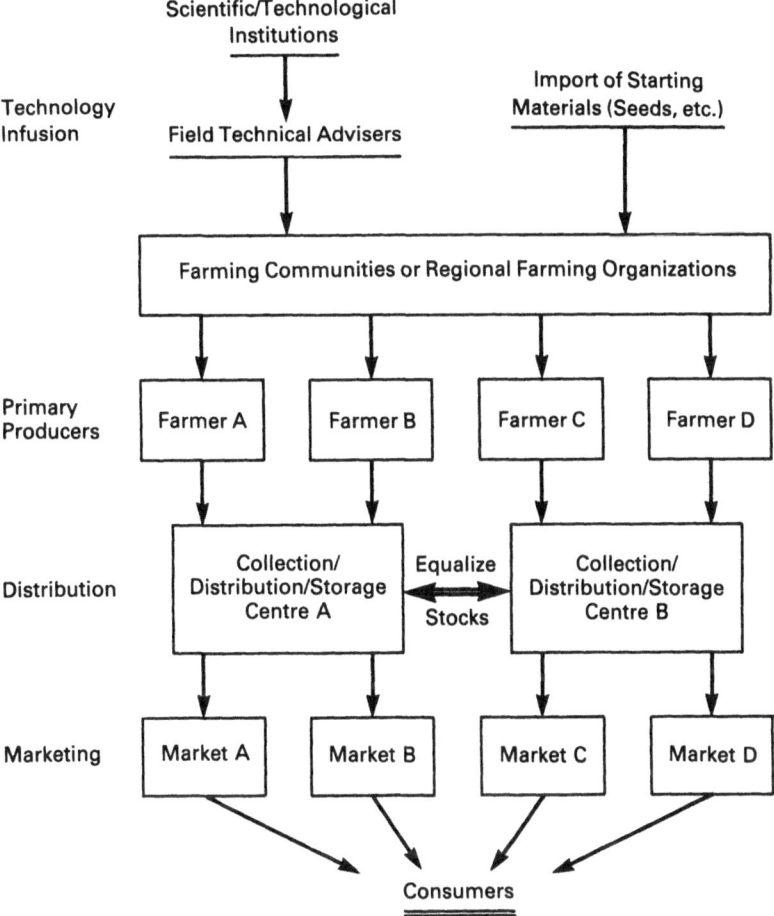

Figure 2.1   Objectives and effect of biotechnology infusion

new careers. Of course, this is to be considered as one possible scenario and not a prediction.

Primary effects

A primary effect will be defined as a change induced on the production or consumption activities of economic agents who first implement a particular bio-product. The primary effects are usually associated with farmers.

Input to output ratios

From an economic point of view, biotechnology results in an improvement of the transformation of inputs into outputs, including improvements in the quality of output. This, of course, means that the input-output ratios must change. Accordingly, the core effect common to all products of agricultural biotechnology is that they are either capable of producing higher output per unit of input or per unit of time. This can be accomplished by genetic manipulation to produce plants or animals with at least one 'output-enhanced' trait. For example, a product with an 'output-enhancing' trait is the so-called rainbow carp produced at the Johns Hopkins University and Auburn University in the United States. The researchers introduced a gene from the rainbow trout into the common carp. The transgenic carp's metabolism was affected, producing a fish that is 20 per cent larger than normal. Another example is the potato cultivar developed in the Netherlands. The plant was made resistant to potato virus X. Given this immunity, we can expect more plants to survive per acre. These examples indicate that enhanced productivity per unit of input can be accomplished with genetic engineering techniques on countless different organisms.

Quality improvement

An illustration of biotechnically-induced changes in input-output ratios, as well as quality improvement, comes from one of the earliest commercial applications of biotechnology in the United States, one frequently cited in the literature. This is the biotechnologically-created natural hormones to regulate animal protein synthesis. These hormones have the ability to both increase the amount of output per unit of input as well as enhance the quality of the product. Use of these biotechnology products will lower feed and time requirements (inputs) necessary to produce a quality product. Another example is that of the genetically-engineered tomatoes which will be vine-ripened, permitting the full flavour to mature without the risk of spoilage in transit.

Cost effectiveness

Decreases in production costs caused by the lower input to output ratios should enhance profitability if they are not accompanied by large increases of industry-wide supply which may lower the market price. Undoubtedly, any production procedure capable of increasing yields significantly while only raising costs slightly, would have tremendous appeal and scope for widespread adoption. The products of agri-biotechnology have the capability of increasing productivity which can be obtained and sustained without significant alteration in the technical and economic infrastructure of a country.

Income generation
Increased output might either raise or lower aggregate farm income depending on each product's price elasticity of demand. However, the production benefits seem directly related to the size of the production: both the small farmer and the large farmer will derive some benefit; the larger the scale, the larger the profit. The significance of this feature of biotechnology cannot be overemphasised. Since most of the farm plots in the Third World countries are not large, the economic consequences of the BR may be more constrained unless there is a reorganisation of farming into more sizeable plots, say into co-operatives. However, this latter change in the production methods may be economically and socially difficult to achieve.

INDIRECT EFFECTS: TECHNO-CULTURAL REPERCUSSIONS

Assessment of biotechnological products

Will the introduction of the products of biotechnology have undesirable social and cultural consequences? Questions about possible injurious repercussions of biotechnology on society have been raised by a number of concerned individuals (Dingell, 1985). In order to reduce the undesirable consequences, the socioeconomic implications of the transfer or development of biotechnology to any developing country needs to be analysed properly (Dohlman *et al.*, 1987). Imported biotechnologies should be examined for cultural appropriateness, responsiveness to people's needs and desires, and consistency with both available resources and the governments' broad social goals. Can the products be assimilated into everyday life without forcing significant cultural and social readjustments?

A note of caution should be sounded, however. Models or systems of technological assessment in developed countries are generally unsuitable for LDCs. For example, the genetically-engineered lean pig might be a highly desirable product where calorie and cholesterol intake are too high, but it might be quite undesirable in certain Third World countries where a high calorie intake is more suitable or where pork is not eaten. Therefore, any analysis of the effects of agricultural biotechnology should take into account the traditional diet of each region. The much praised high-protein products may be incapable of alleviating the basic health problem of nutritional insufficiency if they are not linked to the type of foods customarily consumed. The development of a particular biotechnology product must therefore be culturally harmonised with the regional dietary characteristics in order to arrive at a balanced food intake that includes the required biomass as well as vitamins, minerals, and so on. The risk of unwanted or incompatible biotechnologies may be high if these considerations were ignored.

The engineering of sugar beets that are resistant to specific herbicides provides us with another example of a product that might prove valuable to European sugar beet farmers, who spend over two to three times more on weed control than on seeds. But it is of little value in most Third World countries since cultivation (such as weeding) is done by hand. These kinds of changes may not be advantageous in many LDCs that employ the cheapest and most abundant resource – labour, while conserving the scarce resources – financial capital. Displacement of the labour force may not only induce unemployment, but may also upset the conventional division of labour in the agricultural sector as well as cause imbalances in related job markets. The effect of upsetting the division of labour may precipitate social complications, including gender-specific job dislocations. The techno-cultural effects of the applications of biotechnology are likely to be uneven: they will have beneficial effects in certain sectors while generating negative effects in others.

Agricultural integration of biotechnology products

Existing agricultural products can be modified through genetic engineering to increase their yield without meaningfully altering the properties of the products to which the local population is accustomed. An excellent example of such a product is the Quality Protein Maize (the so-called 'miracle maize') researched and developed by the scientists at Mexico's International Center for the Improvement of Maize and Wheat. Although the maize was not developed by molecular biological techniques, it illustrates what can be expected from biotechnology in the future. This suggests that new variations of agricultural products (that closely resemble older varieties) can be introduced, minimising or reducing the possible social consequences of resistance and disruptions.

Dramatic alterations in farming and ranching techniques due to the introduction of biotechnology may not necessarily be beneficial. In many instances, existing farming techniques can be retained or modified slightly when biotechnical products are introduced to the farmer. Since plants or animals can be engineered for improved characteristics (faster growth, disease and pest resistance, etc.) without significantly altering cultivation requirements including the use of most farm inputs, resistance to adoption should be minimal if the price of the alternative is comparable to local alternatives. Since many of the potential commercial applications of biotechnology to agriculture can be viewed as either resource-saving or output-enhancing, one would expect that economic incentives for adoption may be considerable, even to the point where they overcome cultural inhibitions.

Changes in the scale of farming under biotechnology

According to many experts in farming, several trends that have been occurring in the United States agricultural sector will be accelerated by the use of biotechnology products (US-OTA, 1986). These trends include a reduction in the number of small farms: a trend towards fewer but larger farms throughout the United States. Biotechnology will galvanise a trend not only to consolidation within farming, but also towards a situation where the same multinational enterprises control farming supplies as well as biotechnical research capabilities (Tangley, 1986).

While what happens in the United States may not be applicable to developing countries, disturbing features of these trends do provide certain insights. R & D decisions can place commercial interests above the welfare of the farmer and the general public. On the other hand, the projected huge increases in productivity resulting from the use of biotechnology plants and animals may reduce the need for large numbers of small farms. Such economic efficiency may provide the motivation for LDCs to engage in some sort of consolidation of farms or land holdings.

PROSPECTIVE BIOTECHNOLOGY INDUSTRIES

The capability of genetic engineering to introduce new traits into plants and animals could give rise to new industries which were not foreseen by the pioneers in molecular biology and the founders of modern biology. Unlike the use of biotechnology to develop pharmaceuticals, the application of biotechnology methods to develop plants and animals for consumption as food requires 'low-tech' skills to engage in traditional manual work. These workers form a necessary adjunct to the highly-skilled technicians who are directly involved with the genetic engineering technology. To venture into industries solely dependent on advanced technology may not be the best and most immediate avenue to incorporate biotechnology into the economies of developing countries. Industries that take advantage of abundant 'lesser-skilled' components of the labour force, in combination with advanced biotechnology methods, appear to be the most attractive choice in the agricultural sector. Two of the industries that meet these criteria are the micro-propagation of plants and the embryo engineering of animals. Although their production procedures are vastly different, they share two important factors, that is, similar molecular biological concept and the requirement of a traditional workforce.

Micropropagation industries
The micropropagation (MP) method employs the clonal selection with plant tissue culture that can provide larger numbers of young plants in a shorter period of time than conventional vegetative propagation from seeds, cuttings or grafts. This MP technique is being utilised more and more, not only for ornamental plants, but also for plantation crops and vegetables.

Several advantages of MP over the traditional methods of production can be cited: (1) rapid multiplication of plants with uniform size and characteristics; (2) virtually disease-free plants because of the near optimum and nearly pathogen-free environment; (3) the ability to propagate sensitive species which are difficult to reproduce by conventional means; and (4) the ability to propagate and supply plants in large numbers and on a year-round basis rather than on a seasonal basis.

Micropropagation, especially in combination with genetic engineering technology, has the potential to increase the production of young plants and, at the same time, reduce both the variability and often poor quality of the product. Moreover, the fact that most of the labour inputs are utilised on a continuous basis, rather than seasonally, will stabilise unskilled or semi-skilled employment over time. The number of scientific personnel required is small compared to the unskilled and semi-skilled workers that are needed. For illustrative purposes, the following is a very brief account of the most important productive inputs used by an American firm engaged in the commercial application of micropropagation for about five years. The firm produces about 1.33 million plants annually with about 100 employees. It has a gross revenue income per annum of about US$12 million. However, one needs to bear in mind the considerable variations in revenue and cost across firms depending on such factors as the type of plant propagated, locational factors, energy and labour costs, and so on.

Scientific and technical personnel
If the gene or genes are obtained through technology transfer, several molecular biologists and several experts in plant cell culture will be needed. They will oversee the maintenance of the stock, guide transgenic procedures and possibly manipulate the phenotypes. To assist the above scientists, about ten technicians will be needed to prepare and test the culture medium as well as to conduct routine biochemical manipulations in the laboratory. About 50 additional technicians working on two shifts in sterile laminar flow hoods in the clean rooms will separate, transfer and cultivate the juvenile plants.

Non-technical personnel
About ten 'lesser-skilled' persons will be needed to tend the plants grown in the growth chambers prior to their being moved into the greenhouse. Unskilled workers, about 20 in number, will work in the greenhouse on split shifts, to provide continuous monitoring and care of the young plants to ensure a maximum survival rate and thus a maximum output. Other staff support will be needed for transportation, clerical and office assistance, equipment and building repair, and engineering-related activities. Automation, such as sprinklers for watering plants, could decrease labour input but will increase energy requirements.

Physical facilities
The following is a condensed description of the production facilities (capital requirement) needed to support an MP firm of 100 workers. A building, or buildings, will include laboratories for the R&D, testing and culture media preparation. Clean rooms or operation rooms will house the laminar flow hoods (20–25) for the separation and selection of young plants. Growth chambers should be adjacent or connected to the clean rooms to minimise contamination during transportation. The greenhouse (where the final products will be packaged for shipment) will occupy about 70 to 80 per cent of the total space. Office(s) for business operations, such as a machine shop and garage, can be located within the production facility or elsewhere.

Cost estimates
Table 2.1 summarises the variable and fixed costs based on an annual average for an enterprise in the United States. Since data from LDCs is not available for this kind of an analysis, it is difficult to quantify the comparative costs in LDCs. However, it can certainly be concluded

TABLE 2.1   *Cost estimate for micropropagation*
(millions of US$)

Variable costs

Direct labour and salaries		2.0
Energy (electric and natural gas)		0.2
Water		0.2
Transportation		0.3
Expendable items (culturewares, packing, etc)		0.5
Nutrient mix		0.5
	Subtotal	3.7

Fixed costs

Laboratory		0.5
Operating room (laminar flow hoods, etc.)		0.7
Growth chambers		1.0
Greenhouse		1.0
Business offices and equipment		1.0
Building maintenance		0.3
	Subtotal	4.5
	TOTAL	8.2

that a similar enhancing of overall employment can be expected from a blending of highly-skilled technicians with 'low-tech' skills as described above.

Embryo engineering for animal husbandry

The second illustration of a potential commercial application of biotechnology at the firm level is in the area of embryo engineering. Both the mix and the kinds of capital and labour are dissimilar to that found in the afore-mentioned MP firm. Even though 'new' kinds of animals, including laboratory animals, sheep, pigs, cattle, to name a few, have been successfully engineered, ongoing R&D has not as yet reached a commercially viable production phase for most of these animals. The DuPont Company recently announced that transgenic mice are now commercially available for cancer research. Embryo engineering, therefore, lags somewhat behind MP in actual utilisation by firms making products for the market place. There are few US firms actually engaged in animal husbandry with embryo engineering and they are, for the most part, associated with universities and governmental agencies.

There are reasons for the slow rate of commercialisation of embryo-engineered products. For example, many of these transgenic animals display abnormalities which limit their overall production potential despite the attractiveness of their genetically-engineered attributes. An excellent example to illustrate negative 'side effects' associated with the positive results can be found in some transgenic pigs which have leaner pork, a clearly marketable feature in health-conscious societies in developed countries. Yet these pigs are lethargic, have weak muscles and are susceptible to arthritis and fatal gastric ulcers (*Science*, 1988a). Further reducing the commercial attractiveness of these genetically-engineered pigs are dietary changes in many industrialised countries which strongly suggest a trend away from the ingestion of red meat and red-meat products. This has lowered the expected return from growth-enhanced domesticated animals, somewhat dampening private R & D in this aspect of gene transfer technology. Poultry, especially chickens and fish, would seem to be a more culturally acceptable alternative high-protein source. However, genetic engineering techniques for chick and fish embryos have yet to be developed, although in Chapter 8, the immense potential for boosting human protein nutritional-intake is demonstrated through biotechnology applications to the chicken feed industry in Nigeria.

Developing countries should, however, realise significantly larger marginal benefits from successful advances in embryo engineering which are capable of producing faster-growing and larger livestock animals with the same or less feed requirements compared to normal stock. There are two main reasons for developing countries to begin to adopt embryo engineering technology: first, to increase the quality and quantity of meat and dairy products for domestic consumption, secondly, to expand exports to earn foreign exchange and promote development, although of course extension of current protectionist tendencies may frustrate this prospect. Investment in animal husbandry with genetic engineering technology together with embryological methodology satisfies some of the important preliminary investment-decision criteria – expected benefits versus costs and social norms – to make it a candidate for serious consideration by developing countries.

The following summarises the scientific/technical and non-technical requirements for the establishment of an animal husbandry firm. While the MP firm has been in existence for five years and is profitable, the following animal production unit is a member of a co-operative effort between a university and several venture capital concerns. The hypothetical gross income of US $12 million is a benchmark used to make the figure comparable with the MP firm described above. The variable and fixed costs are formulated according to the actual costs in US dollars.

Personnel

(a)   Scientific (10–20)
Major responsibility will be in R & D of domestic animals, or to maintain and continue the replication of genes. They will have a high food production/feed ratio, a short growth period and have disease-resistant characteristics. Mandatory scientific skills require training in molecular biology, i.e. isolation, production and possible modification of genes and their insertion, via micro-injection or electroporation techniques, into the embryos of mammals and/or poultry to produce transgenic animals. This personnel category will also include veterinarians who will be responsible for the transplantation of cultured embryos into uteri of surrogate mothers and for disease control.

(b)   Technicians (10–20)
Major duties include the preparation of reagents, culture media, specific feed formulas, general laboratory work, clinical testing of animal chemistry, and other routines that require college (or equivalent) training.

(c)   Non-technical personnel for animal care (20–30)
These are farmhands who work in the barns and on ranches. Their occupations are more traditional and require manual labour for cleaning, watering, feeding, herding, etc.

(d)   Business operations (10–20)
These are drivers, office workers, maintenance crew, sales personnel, etc. This category will vary in accordance with regional and business structures.

Physical facilities

Besides the standard office space for business matters, two major R & D and production facilities would be needed. The scientific workers would work in two different kinds of laboratories, a standard one for molecular biological operations and one with an almost sterile environment. In the latter, there would be the carbon dioxide incubators and laminar flow hoods required for transgenic operation and the cultivation of transgenic

TABLE 2.2   *Cost estimate for embryo engineering*
(millions of US$)

Variable cost		
Direct labour		2.5
Utilities		0.2
Transportation		0.4
Expendable items (culturewares, etc.)		0.3
Nutrient mixtures		0.1
Animal feeds and barn supplies		1.5
	Subtotal	4.4
Fixed cost		
Laboratories		0.7
Cultivation and operation rooms		1.0
Barns		0.5
*Pasture lands		1.5
*Vehicles, fencing		0.5
Building maintenance		0.3
Business offices and equipment		5.5
	Subtotal	5.5
	TOTAL	10.0

*   Less requirement during experimental and developmental stages.

embryos prior to transplantation into surrogate mothers. Barns and ranches for animals will occupy most of the space. If poultry is the major output, chicken coops with a strictly-controlled environment would be necessary in order to minimise disease.

Cost estimate

Table 2.2 shows an estimation of costs based on a United States scale in millions of US dollars. The amounts have been adjusted to facilitate a comparison with MP.

INPUT-OUTPUT SIMULATION: A SQUEEZE ON OUTPUT AND EMPLOYMENT

In discussing the potential economic impact of biotechnology, Kalter and Tauer (1987) point out that improvements in animal productivity will have major interactive effects on crop agriculture, an explicit acknowl-edgement of inter-industry linkages within the agricultural sector. They also implicitly recognise that further economic interdependencies connect the agricultural sector with many industries outside agriculture, even in agrarian-based societies such as many developing countries are. Commer-cial applications of the products of agricultural biotechnology are likely to cause significant inter-industry repercussions with possible indirect effects on industries far removed from the agricultural sector. If economic and structural effects are as profound as Kalter and Tauer suggest, then the inter-industry effects will include readjustments in employment and purchases of final and intermediate inputs by industries supplying the agricultural sector.

Less requirement during experimental and developmental stages.

To simulate in a simplistic fashion some of the inter-industry consequences of a resource-saving agricultural bio-product, we utilise input-output tech-niques.[1] Suppose that we consider an M industry model which is the sub-set of all industries (M + N). The M industries are all connected either by utilising or supplying agricultural or agriculturally-related inputs to the primary producers.[2] We are in reality modelling a sub-economy of the aggregate macroeconomy.

Specifically, suppose tissue culture and recombinant DNA techniques develop new crops which are more resistant to disease, drought and/or pests, or alternatively, livestock animals are engineered to metabolise feed more effectively. This could well reduce input requirements per unit of out-put for inputs like irrigation, pesticides, and animal feed. The adoption of a biotechnological innovation of significant resource-saving consequences will generate a series of inter-industry repercussions through the model. The cumulative overall effect of these repercussions, as they spread from one industry to another, may be considerable. The input-output technique is used to analyse the direct and indirect effects on various endogenous industries systematically quantifying the compound impacts. The analysis will also reveal that the exogenous demand for agricultural products may need to adjust to maintain the internal model balance. Figure 2.2 illustrates in a diagrammatic form the flow of goods and services along the industries in the model.

Let X_∞ represent the total amount that industry i (say primary agriculture) must produce to satisfy all demands (from within the industry sub-set or from outside). Let X_{ij} represent the amount of production of each unit of the jth commodity required from agriculture,

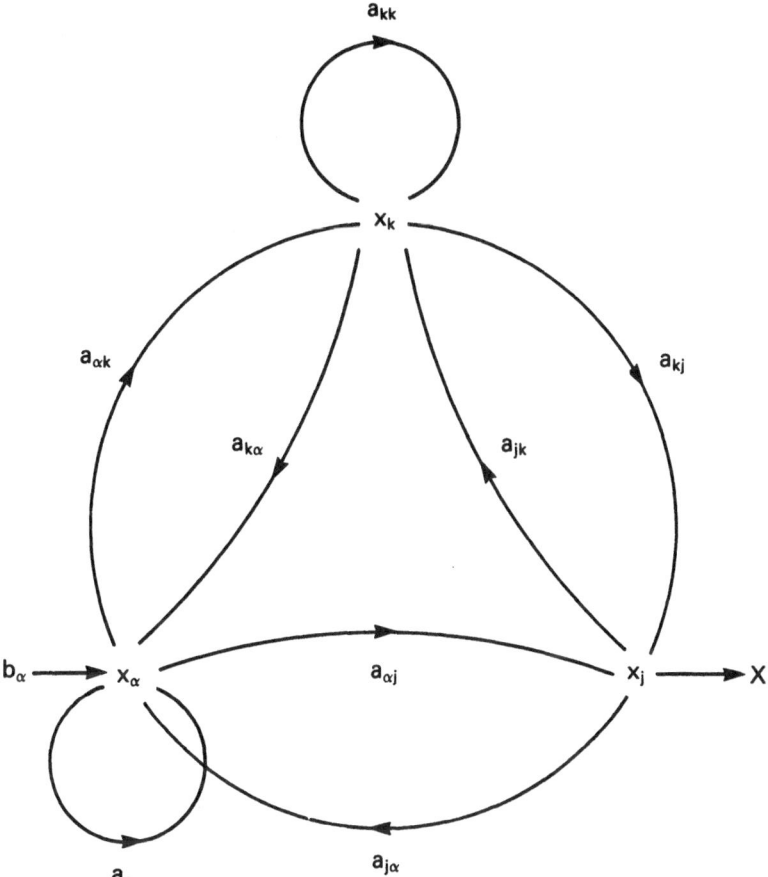

Figure 2.2 Inter-industry flows in the agricultural sector

denoted as the ith industry; ajk is the amount of unit intensity production from the kth industry required by the jth industry, and so on.[3] The a_{ij}'s are the technical input-output coefficients.

In Figure 2.2 we have depicted the technical coefficients and the flow arcs which represent the indirect requirements between primary agriculture and industries k and j. Indirect effects are also evident since the total output of agriculture is dependent on the industries k and j. The incorporation of resource-saving biotechnology products should, over time, reduce the technical coefficients (input requirements) of one or more industries

TABLE 2.3 ' *Variation in gross agricultural output over time under resource-saving biotechnology*

Time period (t)	Gross agricultural output [$x_\alpha (t)$]	Change in exogenous agricultural demand to sustain $x_\alpha(0)$
0	3.173	0.000
1	3.155	0.103
2	3.146	0.020
3	3.130	0.032
4	3.116	0.043
5	3.100	0.054
6	3.080	0.070
7	3.076	0.073
8	3.061	0.085
9	3.041	0.100
10	3.032	0.107

supplying inputs to the agricultural sector. The production technology is altered in the agricultural sector such that:

$$ai \propto (t) \geq i \propto (t+1) \text{ for all } i = 1, 2, \ldots \ldots M$$
$$ai \propto (t) \geq i \propto (t+1) \text{ for some } i = 1, 2, \ldots \ldots M$$

This means that at least one of the input requirements for the agricultural sector is lower in time period t + 1 than in time period t. Accordingly, the simulation generates random decreases in the ai's i=1, 2, . . . M the input flows to the agricultural sector. Some of the results of the simulation are presented in Table 2.3.

Several interesting conclusions can be drawn by looking at the information given in Table 2.3. The results of the simulation indicate that under resource-saving biotechnology the gross output of the agricultural sector will fall unless exogenous demand for output from one or more M industries can be stimulated. The declining gross output in the agricultural sector, shown in column 2, is obtained under the assumption of constant levels of exogenous (autonomous) demand for each of the sub-economies/industries. That is to say if the technical input coefficients fall, gross output in the agricultural sector will also fall. The change in the net requirements for the agricultural sector are felt through the entire economy since this sector directly or indirectly uses products from all other industries.

Column two of Table 2.3 summarises both the direct and indirect impacts on output of a random effect of decline in direct input or output requirements for agriculture. Column 3 indicates the change in exogenous demand for agricultural output alone that would be required to sustain the level of gross agricultural output that occurs in time period 0. Two forces make it more difficult to increase exogenous demand adequately to restore initial gross output X (0). The multiplier effect of increases in exogenous demand on gross output falls over time with the decrease in the direct input requirements per unit of agricultural output. Also, increases in the external demand for other input-supplying industries if they occur would suffer from a similar decreasing impact.[4] The magnifying effect of exogenous increases in demand on gross agricultural output is diluted because the backward linkages to supply industries exhibit smaller flows per unit of output.

Obviously, the tendency for gross agricultural output to decline can be offset by increases in the exogenous demand vector. Less obviously, if additional industries are added to the original M both the direct and indirect effects/coefficients (the matrix multipliers) increase, magnifying the above-mentioned increases in exogenous demand. We can, therefore, conclude that distribution, food processing and transportation and other related industries need to be upgraded, or created, to handle the anticipated surpluses emanating from the output-enhancing biotechnology. This policy suggestion would be appropriate under either resource-saving or productivity-enhancing biotechnology.

The above conclusion, however, is at this stage quite speculative as it is based on a theoretical simulation exercise. This has to be followed-up by verification with concrete empirical data before those conclusions can be definitly confirmed.

CONCLUDING REMARKS

The analysis of this chapter is not summarised, as the discussions on the issues conceptualised in the beginning are extended by means of additional empirical evidence in Chapter 11. Empirical data from China and Kenya (Chapter 5) are used in Chapter 11 to support the conclusions regarding the inter-industry repercussions of biotechnology on GDP and employment. The basic conclusion is that the adoption of biotechnological innovations having significant resource-saving effects will therefore generate a series of intersectoral repercussions throughout the economy. The overall cumulative impact of these repercussions, as they spread from one sector to another, may be very great indeed.

An attempt was made using input-output techniques to trace and quantify these direct and indirect impacts on output (and hence on employment) through a simulation exercise. The results of the simulation reveal that under resource-saving biotechnology the gross domestic product will decline unless exogenous demand for output from one or more industries can be stimulated or new industries are created. Indeed, as has been observed in China, labour released from farming has been deployed in numerous new and biotechnology-related specialised services.

NOTES

1. It would be possible to simulate increased output requirements for supplying industries, but there seems to be little evidence that this will occur under biotechnology. Due to the technical manipulations, please contact the author for a complete description of the specifics of the actual methodology.
2. By decomposing the economy into this subset of industries, the input-output techniques are capable of tracing out economic consequences that might be lost in a more macro-analysis. However, this methodology ignores some potential linkages to other industries in order to remain tractable.
3. The usual assumption of fixed proportions in all production processes is retained within any time period t.
4. In general, the elements in the matrix multiplier $(I - A)^{-1}$ connecting the x_i's with the vector of final demand exhibit declining values under resource-saving biotechnology.

3 New Biotechnologies for Rural Development

Paolo Bifani

INTRODUCTION

Following the broad scenario of the potentials of agricultural biotechnology for developing countries presented in Chapter 2, the purpose of this chapter is to more narrowly focus on microbial biotechnology applications in agriculture. The chapter also deals with its applications in food production and processing. Discussions on the potentials of these biotechnologies and of some selected micropropagation techniques for developing countries are carried out in the context of a broader rural development framework.

POTENTIALITIES OF NEW BIOTECHNOLOGIES FOR DEVELOPING COUNTRIES

Microbial inoculant technology

It has been estimated that at present legumes in association with rhizobia fix at least 35 million tonnes of nitrogen yearly valued at more than US$10 billion. The economic gains from such nitrogen fixation can be seen from the evidence from the United States, in Table 3.1. Nevertheless, legumes obtain only 25 per cent of their nitrogen from nodulisation, while the remaining 75 per cent is obtained from the nitrogen already fixed and existing in the soils in the form of nitrate and ammonium salts. Therefore, at present the limiting factor for leguminous growth is the present 25 per cent of nitrogen fixation through nodulisation. In the United States about 30 million acres of soyabeans are inoculated every year. The annual nitrogen fixation in soyabeans in the United States is estimated at more than 110 kg per hectare valued at about US$375 million. The advantages of inoculation are: it requires little capital, the transport costs are minimal and the technique itself is relatively simple and can be easily applied in rural areas of developing countries. Furthermore, it can be done in accordance with the local requirements and can also generate additional jobs.

TABLE 3.1   *Nitrogen fixed by rhizobium legume association*

Crop	Nitrogen fixed (kg/ha/year)	Estimated value as fertiliser at 1975 prices (US$/ha)
Alfafa	128–300	38–90
Clover	118–155	35–47
Cowpea	84	25
Faba bean	240–325	72–98
Lentils	103	31
Peanuts	47	14
Soyabeans	60–80	18–24

SOURCE: National Academy of Sciences, 1979.

Rhizobial inoculant in Brazil

The state-owned Empresa Brasilera de Pesquisa Agropecuaria (EMBRAPA) initiated the process of identification of indigenous strains and their reproduction for the production of inoculants. The purpose was to replace imported chemical fertiliser that represented about 75 per cent of the total cost of production of soya. The experiment has been successful and already in 1981 the 4.5×10^6 hectares under soya cultivation use only locally-produced rhizobia inoculant. It is estimated that in 1981 the use of rhizobia inoculant in soya cultivation saved the equivalent of US$1000 million of chemical fertilisers. Furthermore, the soyabean crop has been used to expand the agricultural frontier by the incorporation into agriculture of edaphic savannas called *cerrados*.

A particular problem is the seasonal aspect of the inoculant production which results in long idle periods of the fermentation equipment. This problem can be partially solved by the expansion of production to other inoculants. The Brazilian inoculant market is potentially very large since it is not limited to soya but includes also the black bean (*feijão*, which is the main staple of the Brazilian diet), peanut, legume-based pastures, including the legumes for under-used savannahs, forestry, etc. The Brazilian production of *feijão* is of about 2 million tonnes per year on an area of 5.3 million hectares and the production of peanuts is about 300 000 tonnes per year on an area of 220 000 hectares (Zylberstajn *et al.*, 1985).

It is important to note that the use of inoculant saves capital but not labour. This is for two reasons: first, the inoculant replaces nitrogen but no other fertiliser and secondly, the seed of the leguminous crop should be coated with the inoculant; this is a task that, in Brazil, is done practically on the field and is little mechanised. Furthermore, when the legume culture is used to fix nitrogen for a following crop that has no fixing ability, say for example, cereals, or whenever a mixed-cropping system is practised, the switch to mechanisation creates problems of labour displacement. Therefore the net impact of the inoculant technology is likely to be more positive in terms of labour requirements. This impact on labour does not include the consideration of the creation of new, although limited demand for skilled labour in the manufacture of the inoculant. The operation of the inoculant plant requires professional and well-trained staff to work on fermentation processes.

Other developing countries (e.g. Argentina, Uruguay and India) have also made successful experiments in the production and use of rhizobial inoculant. In India, seed inoculation with rhizobium in rain-fed conditions has resulted in average increases of yields of 27 per cent for chickpea, 26 per cent for pigeon pea and 38 per cent for mung bean.

A second natural process of fixing nitrogen through symbiosis is between an organism called frankia and non-leguminous plants, mainly woody shrubs and trees. At present about 145 species of plants of this type are known. Among the most important of these plants are the casuarina. In Senegal soil around casuarinas, the nitrogen increased at rates of about 60 kg per hectare.

Labour-intensive symbiotic nitrogen fixation: China, Egypt and India

The third natural process of symbiotic fixation of nitrogen occurs by the association of the freshwater fern called azolla and a blue green algae (BGA) known as anabaena. This association is one of the most efficient processes of nitrogen fixation from the energy point of view. The azolla-anabaena symbiosis can fix between 100 and 150 kg of nitrogen per hectare per year in approximately 40 to 60 tonnes of biomass, if multiple crop growth cycle of the association is practised over a year (Alaa el Din *et al.*, 1984). This amount of biomass can be achieved because, under favourable conditions, azolla vegetation can double in about five days. However, the normal amount fixed is between 50 and 75 kg/hectare per crop and can be achieved with an active growth period of six weeks.

Dead azolla decomposes in the soil releasing 56 per cent of the nitrogen compound in three weeks which then becomes available to the plant, normally rice. One crop of azolla applied to paddy rice is equivalent to the application of 30 kg/$N2$ per hectare of urea (see Table 3.2).

An additional advantage of the azolla derives from its high protein content. For this reason, it represents an important source of feed. In China azolla constituted up to 50 per cent in the diet of pigs and one

TABLE 3.2   *Effects on grain yields of azolla incorporation
compared with application of chemical nitrogen fertilisers*

Treatment	Grain yield (1979)	Tonne/ha (index) (1980)
No nitrogen	2.6 (100)	3.2 (100)
Nitrogen (30 kg/ha)	3.2 (122)	3.8 (123)
Nitrogen (60 kg/ha)	3.7 (141)	4.2 (139)
Azolla growth before transplanting	3.2 (122)	4.0 (123)
Azolla after transplanting and incorporation	3.1 (118)	3.9 (123)
Azolla after transplanting but not incorporated	3.1 (119)	4.0 (123)
Nitrogen (30kg/ha) plus azolla before transplanting	3.7 (143)	4.4 (140)
Nitrogen (30 kgh/a) plus azolla after transplanting	3.5 (134)	4.4 (140)
Azolla before and after transplanting	3.6 (139)	4.2 (137)
Number of sites	13	19

SOURCE: Swaminathan, 1982.

hectare of green azolla provides enough fodder for 200 pigs (FAO, 1977 and 1978).

Inoculation of paddy rice with blue-green algae is widely used in Egypt and India where blue-green algae can contribute up to 77 kg/ha/per cropping season. In the USSR, inoculation with BGA resulted in increases of rice yields by 13–20 per cent while in China a 24 per cent yield increase was achieved. The studies carried out in Egypt and India demonstrated that the inoculation with BGA is equivalent to the application of 24–48 kg/N/hectare, with the additional advantage that algalisation leads to increases in the organic matter component of the soil, hence augmenting its fertility. As in the case of the rhizobium, the BGA presents some local characteristics. The BGA used for the inoculation should be from the same area where it will be applied. The inoculation of paddy rice with BGA was initiated in Egypt during the 1950s; it was planned to inoculate 21 000 hectare in 1985 and to reach 420 000 hectares by 1990. The inoculated fields gave 26 per cent higher yields than those without inoculation. This process resulted in increasing returns to the Government which produces the inoculants for selling to the farmers. Farmers obtain higher returns due to a reduction in the use of chemical fertiliser from an increasing yield obtained.

In India the Agricultural Research Institute provides to the peasant starter culture of BGA. It was calculated that from an area of 25 m^2 it is possible to obtain, every 15 days, about 100kg of algae that can be used in rice paddy. The inoculation of 10kg/hectare of BGA supplemented with phosphate results in a benefit of between 20 to 30 kg nitrogen/hectare per season. The response in terms of crop productivity has been positive,

TABLE 3.3 *Economic returns to the Government and rice farmer in Egypt*

Economic value	Fertilisation by mineral-N¹ (40 kg N/feddan)²		Fertilisation by 100 gm BGA and 20 kg mineral-N/feddan		Total economy
	Tonne/ feddan	Egyptian £/ feddan	Tonne/ feddan	Egyptian £/ feddan	
A. *For the country*					
Value of the mineral-N units	–	16.200	–	8.100	–
Value of algae inoculant	–	–	–	1.000	–
Total cost of fertilisation	–	16.200	–	9.100	–
Economic return from fertilisation	–	–	–	–	7.100
Average of total yield (paddy)	2.549	–	3.218	–	–
Yield of seeds (white rice)	1.784	–	2.252	–	–
Price paid to the producer	–	151.640	–	191.420	–
Selling price to the consumer	–	249.670	–	315.280	–
Economic return from the yield	–	89.120	–	123.860	25.740
Total economic return	–	–	–	–	32.840

TABLE 3.3 Continued

Economic value	Fertilisation by mineral-N¹ (40 kg N/feddan)²		Fertilisation by 100 gm BGA and 20 kg mineral-N/feddan		Total economy
	Tonne/feddan	Egyptian £/feddan	Tonne/feddan	Egyptian £/feddan	
B. For the farmer					
Value of mineral-N units	–	10.360	–	5.180	–
Value of algae inoculant	–	–	–	1.000	–
Total cost of fertilisation	–	10.360	–	6.180	–
Economic return from fertilisation	–	–	–	–	4.180
Average of total yield (paddy)	2.549	–	3.218	–	–
Yield of seeds (white rice)	1.784	–	2.252	–	–
Average of selling price	–	167.981	–	212.048	–
Economic return from the yield	–	–	–	–	44.067
Total economic return	–	–	–	–	48.247

1 Chemical nitrogenous fertiliser.
2 1 feddan = 0.42 hectares.

SOURCE: Alaa El-Din and Shalan, 1984.

New Biotechnologies for Rural Development 49

TABLE 3.4   Potential benefits of inoculation with azospirillum of sorghum,
pearl millet, finger millet and barley in India
(all India coordinated field trials)

Crop	Yield increase (per cent)	Saving in nitrogen (gN/ha)
Sorghum	9 to 30	40
Pearl millet	4 to 83	13 to 20
Finger millet	2 to 31	13 to 20
Barley	15 to 26	40

SOURCE: Subba Rao, 1984.

reflecting increases between 9.3 per cent and 11 per cent in the states
of Madhya Pradesh and Tamil Nadu respectively. Field experiments in
India reveal very encouraging results, as is shown in Table 3.4, and an
additional advantage of the BGA is that it can be stored in dry form
without losing its reproductive capacity, so that the dry material can
be used at a later date directly in the paddy fields or for new algae
production.

The economic impact of this short-term strategy in developing countries would be considerable. The potentiality of the Brasilian microbial
inoculant market has already been noted. In Argentina, the potential
market for rhizobial inoculant is estimated at US$10 million. Finally,
the potentialities of Indian agriculture to absorb microbial fertiliser is
illustrated in Table 3.5.

Biopesticides in Brazil

Brazil has experience in the production of microbial insecticides and
controlling pests that affected two of the most important crops: soyabean
and sugarcane. A major Brazilian problem on soyabean plantation has been
an insect responsible for as much as 40 per cent of losses due to insects
(the anticarsia gemmatalis known in Brazil as 'largarta de soya'). In 1972,
a virus of the baculovirus family was isolated at the Centre Nacional
de Pesquisa da Soja (CNPS) laboratory or the governmentally-owned
EMBRAPA. In 1983–84, the virus was successfully applied in more
than 11 000 hectares of the states of Parana and Rio Grande do Sul.
This acreage was increased to 300 000 in 1984–85 and the cost of the
protection in soyabean cultures of these states was reduced by 75 per cent in
relation to the traditional chemical control (Moscardi, 1983). By 1985, five
co-operative schemes were created in Parana, one in Santa Catarina, five in
Rio Grande do Sul and one in São Paulo, all of them with the institutional

TABLE 3.5   *Requirements, present capacities and shortfalls in annual microbial inoculant production in India*

Crop/microbial	Total acreage (ha ×10⁶)	Minimum recommended rate of microbial fertiliser application (kg/ha⁻¹)	Total requirements (kg ×10⁶)	Current estimated (kg ×10⁶)
Grain legume/rhizobial inoculation	30	0.5	15.0	0.8
Pearl millet/azospirillum inoculation	16	0.5	4×10	nil
Sorghum/azospirillum inoculation	16	0.5	8.0	nil
Rice/blue green algae	40	10.0	400.0	0.04

SOURCE: Subba Rao, 1984.

and scientific support of EMBRAPA/CNPS for the commercial production of the virus and their diffusion.

The second experience in the production of viral insecticides refers to the control of the 'diatraea saccharalis', a major problem on the sugarcane plantation. It was estimated that a plague with an average index of infestation of only 2.5 per cent can result in loss of US$100 per hectare of cultivated sugarcane. In 1978, a particular virus (DsGV) was isolated in the genetic department of the Universidad de Campinas (UNICAMP), near São Paulo, and its virulence increased by a factor of 100 through genetic engineering. It has been estimated that the cost of the application will be about US$10 per hectare. The production is at present in a scaling-up phase (Barros, undated).

Micropropagation of cassava

Cassava production offers an interesting area for biotechnology application in developing countries along the complete food chain. Biotechnological techniques can be applied in the inoculation of micorrhiza, for the detection of diseases, production of disease-free plant material through *in-vitro* cloning, for the biological pest control, for the processing of cassava for industrial purposes, and in particular its conversion to alcohol or macrobial proteins through SCP systems. Finally, they can be used for the upgrading of traditional food processing based on fermentation like gari and foo-foo.

TABLE 3.6 *Yields of cassava in Colombia after the eradication of the Caribbean Mosaic through tissue culture and comparison with an hybrid variety*

Clone	Cycle	Fresh Roots (Tonne/ha)	Starch (Tonne/ha)	Shoot-tip No. of plants
Secundina	1st year	25.1	7.1	10
	2nd year	23.0	6.8	10
	3rd year	22.0	5.6	9
	control	8.9	2.1	3
CM.342.170	1st year	34.8	7.9	14
	2nd year	36.2	8.4	10
	3rd year	15.1	3.1	6

SOURCE: Roca, 1985.

In vitro clonal propagation has been applied to eliminate crop disease in the case of potato and cassava at CIAT in Colombia. In 1972 the frog-skin disease destroyed 7000 hectares of land cultivated with cassava crops at Quilcacé, Cauca, Colombia. After the disease the production per hectare fell from 15 tonnes to only 0.5 tonnes (CIAT, 1985). As a consequence the crop was slowly abandoned and the area cultivated was reduced to only 500 hectares. The disease was transmitted through infected shoot-tip and tools used to cut infected plants.

The propagation of disease-free clones of a variety called secundina, traditionally affected by Caribbean Mosaic disease, endemic in the north of Colombia, increased from 9 tonnes per hectare to 25 tonnes per hectare and the production stabilised after three years, while the yield of hybrid varieties, in spite of higher increase in the first two years, declined in the third one (see Table 3.6) because of its lower resistance to diseases. The yield in fresh weight and starch content increased by 70 per cent (Bifani, 1988). The application of the *in-vitro* method to clean cassava infected with frog-skin disease in Quilcace resulted in planting materials that, after one year of distribution to the farmers, show increases in yield from 0.5 tonnes/hectare to 10 tonnes/hectare, recovering from the crisis of 1972, while the acreage allocated to cassava jumped to 5000 hectares. Abandoned treatment plants were replaced and production of starch resumed. However, as a consequence of the increased cassava production, its price is now falling.

In vitro propagation to cassava was also applied to a variety called 'Llanera' which, showed continuously declining yields. The study was done in association with another variety, the CMC-40, characterised by its good performance. Thermotherapy application followed by *in vitro*

TABLE 3.7   *Yield increases from* in vitro *propagation of cassava*

Culture	Parameter	Yield increase (%) over control	
		1st year	2nd year
Llanera	Root-yield – fresh weight	53	25
	– dry weight	22	10
	Planting material	47	46
	Plant height	25	0
	Height first branching	76	26
CMC-40	Root-yield – fresh weight	43	29
	– dry weight	29	2
	Planting material	18	23
	Planting height	6	5
	Height first branching	19	14

SOURCE: Roca, *op. cit.*

propagation resulted in increased yields of roots and planting material in both varieties, in particular during the first year (see Table 3.7). The utilisation of *in vitro* propagation in other, apparently healthy cassava varieties, resulted in yield increases of 20 to 30 per cent.

In Nigeria, more than 80 000 hectares have been planted with new cassava varieties with higher yields ranging from 50 to 300 per cent, more stable production because of greater resistance to pests and diseases, reduced weeding and improved quality. Similar results have been achieved in other countries: In Sierra Leone, the new IITA varieties show higher yields ranging from 9.7 to 25.5 tonnes/hectare, i.e. two to four times the yield of local varieties. In Gabon, in 1982, the new varieties produced high yield ranging from 20 to 40 tonnes/hectare, while in the United Republic of Tanzania, the average yields are 30–35 tonnes/hectare, which compared vary favourably with the 15 tonnes/hectare yield of the local varieties. With the new varieties cloned at IITA, yields ranging from 20 to 50 tonnes/hectare have been achieved in Rwanda, Liberia, Seychelles and Cameroon.

Micropropagation of palm

Nearly 90 per cent of palm oil is used as edible oils and fats, frying oils and fats, non-dairy creams, margarine, cocoa butter substitutes, bakery and pastry fats. The remaining 10 per cent palm oil used in non-edible product goes for the production of soaps, resins, stearic acids, glycerine, detergents or feedstocks for the petrochemicals industry. The diversity of its uses is undoubtedly one of the reasons for the increasing production and trade. By the middle of the 1980s, the total world exports of palm oil represented around 25 per cent of the total exports of vegetable oils and fats which fluctuate around 19 million tonnes.

The international market for palm oil is under the control of the largest multinational enterprise of the food sector, Unilever, which indeed controls margarine, oil, detergent and soap markets. The main policy of Unilever has been to acquire the control of the raw materials. Unilever has about 90 000 hectares of plantations in developing countries, two thirds of which are oil palm trees. In order to face the increasing demand for oils of more than 3 per cent per year for the last two decades. In 1968 Unilever initiated tissue culture technology for the improvement of Malaysian palm plantations. In 1977 the first clones were planted at Unipamol Kluang. The cloning of palm oil should allow Unilever to overcome the problem of hybridisation and at the same time, the elimination of variability resulting from the sexual reproduction. More uniform plants permit easier harvesting, reaping at the same time oil of more uniform characteristics and a 30 per cent increase in production in normal average climatic and environmental conditions. This increase in production will more than compensate for the high cost of *in vitro* propagation, which has been estimated to be between five and ten times the cost of the seeds.

In 1984 UNIFIELD which is a Unilever joint venture with Harrison and Crosfield, produced 200 000 plantlets in a plant near Bedford in the United Kingdom. A similar unit is planned in Malaysia. The target is the production of 1 000 000 plants per year. However, Unilever ambitions are not limited to the short term. Through molecular genetics the idea of the ongoing research is to find out the better varieties by the insertion of genes on plant cells as they are growing on tissue culture. The cross-breed selection at molecular level through the recombinant DNA techniques is the long-term strategy pursued by Unilever to keep control of the world market of detergents and oils. In 1982, Unilever R & D expenditures were estimated at US$344 million in laboratories in several countries. In the very long term, the countries producing palm oil will face the risk of its being obtained from bacteria, which could eliminate them from the market. At present, it is calculated that one tonne of equivalent palm oil obtained from bacteria could cost around US$3000.

In 1984, tissue culture R & D is estimated to have cost Unilever more than US$2.6 millions. The clones produced are the subject of field trials in Brazil, Colombia, Indonesia, Zaire and Papua New Guinea.

Clonal propagation of oil palm tree has also been undertaken by the

French Institut de Recherches pour l'Huile Oleagineuse (IRHO), which in 1976 obtained new plantlets by *in vitro* culture of leaf tissue of young leaves. The process developed by IRHO and Orstom considers two phases for the micro propagation. The complete process lasts around 16 months. The first trials of clones obtained through this process were initiated in 1983 in Côte d'Ivoire. A production of 50 000 *in vitro* plants per year was reached in 1984. IRHO and Orstom have signed contracts with the Federal and Land Development Authority (FELDA) of Malaysia and SOCFINDO, a plantation company in Indonesia, for the joint propagation of clones of oil palm trees.

The biotechnology applications in palm oil is expected to help to overcome biological constraints. The African palm oil appears to be susceptible to South American parasites, thus limiting their expansion. The other objective is the production of a variety more resistant to drought, which is a limiting factor in Africa. Apparently, field trials in Benin, Cameroon, Colombia and Peru have led to positive results.

More on tissue culture

Coconut trees face problems similar to oil palm trees, e.g. the long time generation of seed propagation and high variation between individual plants. In addition, the different characters of coconut trees manifest themselves after ten years. The cross breeding could therefore take several decades. It is estimated that high-yielding cloned coconut trees can produce five times more output than the traditional trees.

In horticulture, tissue culture has witnessed important developments in several developing countries. This contributes to the diversification of production and introduces possibilities of a new agricultural export activity. In Brazil, the EMBRAPA initiated, at its *Centro de Investigaciones de Frutales de Clima Templado*, a large programme for the production of fruit typical of temperate areas in order to compete in the international markets. In 1983 the tissue culture of strawberry meristem reached a production of 20 million plants of six different cultivars (one for industrial process and five for direct consumption). The productivity rose from the Brazilian average of 3 tonnes per hectare to an average of 12 tonnes per hectare with high maximum values of 22 tonnes per hectare. This production represented 70 per cent of the total Brazilian mechanised plantation of strawberries (Bifani, 1988). This success has encouraged EMBRAPA to expand its programme so that in 1984, with the same technique, it was possible to produce 10 000 units of mulberry plants which are being distributed to private producers for commercial development.

Another important development concerns potatoes. In Argentina, the micro-propagation of healthy plants enabled the country to become self-sufficient in only five years. In 1980, Argentina imported 41 million tonnes of potato material for seed at a cost of US$14 million, and in 1985, the import went down to only 1560 kg for a total cost of US$807. In Brazil, EMBRAPA is producing about 40 000 tubers of four different disease-free cultivars of potato. In 1985, it initiated the commercial micro-propagation. Important results have been achieved in ornamental plants in Brazil, Venezuela and other countries.

FOOD PRODUCTION AND PROCESSING

The food problem and biotechnology for protein production

Three possible strategies (or a combination of them) can be considered for the improvement of protein production and consumption in developing countries, through the utilisation of biotechnologies. The first one concerns the improvement of existing fermentation processes most frequently used in different regions. The other possibilities are based on the fact that micro-organisms are very effective in protein conversion and that they can grow in different substrates, constituting finally a product in itself. Thus, production of microbial protein is the main element of the other two strategies for feeding livestock or, as in the third strategy, for direct consumption by humans.

Improvement of traditional fermentation (food) processes

In many cases biotechnologies will make it possible to upgrade traditional technologies in developing countries. In Indonesia the fermentation processes for the preparation of *tempeh* and *tape ketella* have been a practice for centuries. This practice has allowed an increase in nutritional value of food which is within the economic reach of the rural population. *Tempeh* not only increases the protein content of the soyabean substrate, but it also introduces vitamin B12 in the final product.

Similarly, through fermentation of the *tape ketan* the lysine content of rice is increased by 15 per cent and the thiamine content, three times while the protein content of rice is doubled. The *tape ketella* fermentation permits an increase in the protein content of cassava, between two and four times its original content.

Microbial protein production

A more ambitious strategy is the large-scale and maximum production of one microbial protein or a range of different microbial proteins for animal feed and human food. Microbial SCP is the bio-technological process to achieve this objective. SCP is defined as dried cells of micro-organisms such as algae, actinomycetes, bacteria, yeast, moulds and higher fungi, which are grown in large-scale culture systems for use as protein sources in human and animal feeds (Litchfield, 1984).

The process takes advantage of the efficiency of micro-organisms for the production of protein, which is reflected by the fact they can develop between 100 and 1000 times more quickly than a plant or an animal. This advantage is coupled with the fact that these micro-organisms can reach a high degree of protein content – between 40 and 80 per cent on a dry weight base thus constituting a product in themselves.

The following are the advantages of SCP production:

– production is relatively free from seasonal variations and from hazards inherent in ecological climate interaction;
– production can be undertaken using a great variety of substrates; thus it can be adapted to the conditions and endowment of resources of different regions;
– short generation time and reproduction rate of microbes (one to two hours) permits a rapid increase of biomass and protein production;
– production of microbes can be enhanced by the selection and genetic manipulation of different strains for the achievement of optimum growth rate and aminoacid compositions.

SCP production in Mexico and Cuba

Pilot project developments have been undertaken for SCP production at *Cinvestav* in the *Instituto Politécnico Nacional de México*. Obtaining microbial protein from sugarcane bagasse forms part of a larger attempt to obtain protein from cellulosic waste, which includes barley, straw, cornstover, and so on. The implication of this is that a process for the pre-treatment of lignin should be undertaken before the fermentation process.

In the Mexican case, an alkaline method has been adopted. The mixed bacterial culture is then grown on a substrate of detrited bagasse and mineral salts. At the end of the fermentation cycle around 88.5 per cent of the fermented broad is retired, containing about 43.5 g/lt of bacterial cells. The bacterial cell cream is separated from the residual substrate, dried and stored. The efficiency of the recovery is calculated at 98 per cent. The product is commercialised in the form of dry powder. In addition, the biomass suspension that has been separated from the protein cream is floculated and concentrated by centrifugation, obtaining a dry biomass with 12 to 15 per cent of protein that can be used as forage with the same characteristics of dry alfalfa.

The total capital investment is around US$78 million for a plant of 50 000 tonnes capacity per year. The critical point of the process is the pre-treatment with alkali. If the consumption of caustic-soda, which at present represents almost 35 per cent of the cost, can be reduced, the economic possibilities of the process will be enhanced. At the moment Mexican experts are examining the process developed at CENIC in Cuba which is based on the treatment of bagasse in a solid substrate fermentation process. It results in lower chemical and energy requirements, lower capital investment and reduced generation of wastes, with a higher substrate fraction available for micro-organisms.

Another possibility envisaged for the reduction of the investment and overall costs is the shortening of doubling time of the biomass. The CINVESTAV project assessment was done on the basis of a doubling time of 5.8 hours. However, the laboratory studies at the IPN indicate that is is possible to achieve a doubling time of 4.5 which can result in capital reduction of about 10 per cent. Therefore, the experts of the CINVESTAV-IPN conclude that a reduction of about 60 per cent in alkali consumption and 10 per cent in capital investment are possible. Therefore, the SCP protein can be marketed at a price lower than US$1000 per tonne. Further economic reduction can be achieved if the hemicellulose and lignin can be recovered.

In Cuba, important developments have been made to utilise molasses for the production of proteins by using the waste from distilleries. The development of microbial proteins was stimulated by the existence of large quantities of can molasses and the need to reduce imports of cereals for animal protein production. The Instituto Cubano de Investigaciones sobre los Derivados de la Caña de Azúcar decided to redesign the plant, in accordance with Cuban conditions. Since then around 10 plants have been built. The production of these plants is oriented to animal feed, in particular pork and beef production (Bifani, 1988).

The process adopted in Mexico for the processing of SCP from molasses is based on the plants already operating in Cuba, where the molasses are pre-treated in order to eliminate impurities and subsequent sterilisation for the fermentation process.

Enzyme engineering

Another important use of micro-organisms in food processing refers to the production and utilisation of large molecules, like isolated enzymes and plant, animal or bacterial systems as catalysts in biological transformation processes. The most important products of these micro-organisms are the enzymes.

Traditionally, enzymes were obtained from vegetables and animals as a by-product of slaughter houses. These developments make the microbial production of enzymes more attractive than the traditional ones, which present several limitations.

Today microbial production of enzymes by fermentation using fungi, bacteria or yeasts has largely replaced the plant and animal enzymes. Microbial production of enzymes is not seasonal since it is obtained in artificial sterilised cultures, and since it is not produced as a by-product, it does not depend on the vagaries of the main product.

Isoglucose or High Fructose Corn Syrup (HFCS)

From an economic point of view, the major impact of immobilised enzymes has been in the sugar sector and in particular for the production of high fructuose corn syrup (HFCS) or isoglucose which is the first and most important large-scale industrial application of immobilised enzymes. It took only ten years for isoglucose to acquire the control of 10 per cent of calorific sweeteners in the United States. Isoglucose recently accounted for 28 per cent of the North American sweeteners' market and 45 per cent of the total industrial sugar market; both shares are much higher at present. In Japan, isoglucose covers 10 per cent of the sugar demand.

The isoglucose permits a better conservation of food through slower development of bacteria, without the need for additives. At the same time, it lowers the freezing point, which makes it very attractive for ice-cream preparation.

Corn represents 50 per cent of the cost of isoglucose and energy 20 per cent, while between 10 and 43 per cent is the labour cost and only 5 per cent is the cost of the enzymes. It is expected that isoglucose will substitute for up to 90 per cent of sugar in the beverage sector, 60 per cent in canning, 40 per cent in processed food, 30 per cent in dairy production and 25 per cent in baking. This rate of substitution is important since liquid sugar accounted for 57 per cent of the total sugar use in the United States. The main consequence of the development of isoglucose has been the reduction of sugar imports from developing countries, also discussed in detail in Chapters 7 and 11. The main developing countries which have been affected by this are: the Dominican Republic, with a reduction in exports of 39 per cent, El Salvador, 72 per cent, Nicaragua, 46 per cent and Argentina and Brazil, 34 and 35 per cent respectively.

Crott indicated that the American market absorbed 80 per cent of all the sugar exports of the Dominican Republic, which represents 25 to 40 per cent of total export revenues, or about 60 per cent of foreign exchange earnings of the country. The development of isoglucose in Canada has affected mainly Australia, South Africa and Cuba, which account for almost 90 per cent of all raw sugar imported by Canada. In this case, the most affected countries have been the marginal suppliers, those responsible for the remaining 10 to 20 per cent, particularly Jamaica, Guyana, Fiji, Thailand and Mauritius.

CONCLUDING REMARKS: THE SCENARIO

New biotechnology should reduce the dependence of the developed countries on imported vegetable proteins and other more specific products like medicinal castor oil, citronella, palm oil or cocoa butter. SCP from fossil fuel or from agro-industrial residues like cellulose, molasses, pulp and paper effluent or whey from dairy industries is likely to displace the soya meal protein produced in Argentina and Brazil, the cassava pellets produced in the Philippines or the peanut oils from western Africa.

Moreover, biotechnologies are blurring sectoral boundaries by linking apparently totally different products, through price fluctuations in the international markets. Actually the diffusion of HFCS and SCP has been governed by the changes in the price relationship between apparently non-linked feedstocks and commodities: oil, sugar, maize, molasses, soyabean meal and so on (see chapters 7, 8 and 11). Therefore unexpected changes in the international market may well deter or postpone the diffusion of biotechnological innovations, particularly when international price fluctuations affect raw materials. The case in point is the one concerning SCP. The low price of oil during the 1960s stimulated the development of SCP protein on the basis of methanol and paraffin as substrate.

The Cuban production of SCP protein was made possible by the regular supply of a byproduct of high protein content, molasses, with a very low opportunity cost, that can be used as a cheap raw material in a process in which the incidence of raw material cost is important (around 60 per cent). Actually the cost of production of molasses is almost nil. Moreover, the declining price of sugar in international markets contributed to keeping the opportunity cost of the byproduct at a very low level, thus stimulating its use.

In Cuba, SCP has totally replaced the import of protein for cattle feed. It creates a new sort of employment and increases the value of sugarcane plantations since molasses are used as subtrate. However, the creation of employment by plant activities is relatively low. In the case of Cuba, it is slightly more than 100 persons. But the indirect employment effects can be important, e.g. through the use of additional land and the release of foreign exchange through discontinuation of imports of animal feed.

One of the case studies mentioned in this chapter concerns the increasing application of rhizobium inoculants for synthetic fertilisers in soyabean production in Brazil. The direct employment effect, on which we do not have data, can be important, since the soya expansion in Brazil in the last ten years was undertaken to open the agricultural frontier. The production of rhizobium has a positive impact on rural employment and incomes, and it can help save foreign exchange currently spent on imported fertilisers. Similar considerations are valid for the production and diffusion of microbial pesticides.

In food processing, the application of biotechnology is not likely to have important direct employment effects. Normally, fermentation activities absorb little skilled labour. Furthermore, the tendency is towards continuous processing which has the advantage of raising productivity but a disadvantage in reducing demand for labour. The direct employment impact is likely to be further reduced if automation is applied. For example, the production of glucose isomerase is computer controlled. Therefore, the most important impact is indirect, mainly accruing from the production, pre-processing and transport of the raw material. For example, for SCP the major component of the cost is molasses or bagasse and in the case of HFCS, it is maize. In both cases the organic substrate represents over 50 per cent of the cost of production.

Qualitatively, in developing countries the applications of new biotechnology in agriculture and food processing are likely to result in demand for more skilled labour. They also call for a more flexible occupational structure.

The production of inoculants requires local capabilities for the identification of the most appropriate rhizobium strain and its isolation and reproduction. Besides, during the manipulation of the inoculants, certain skills would be needed to preserve the effectiveness of the inoculant and to coat it appropriately with the seed. The tissue culture process, too, requires skilled persons, both at the laboratory stage and at the field.

The existing views are split regarding the employment and income implications of new biotechnology. Some people claim that the diffusion of new biotechnology will displace labour and reduce income opportunities, depressing the already depressed rural areas of the Third World. At the other extreme are those who believe that the new biotechnologies will create new activities and therefore new jobs and sources of income. The advocates of this position go further by arguing that biotechnology diffusion will require specific services and infrastructure which would have a favourable multiplier effect on the overall development in the Third World.

An important question is the time horizon of the impacts of the new biotechnological developments. In previous paragraphs we show that their diffusion is governed to a great extent by economic and political decisions tailored to serve specific purposes. On the other hand, we also noted that several new biotechnological innovations are not likely to make an early entry in the productive process. Furthermore, since diffusion is a rather slow process the social impact will manifest itself after a rather long period. For example, tissue culture in tropical plants like palm oil or coconut is not likely to yield impacts in the near future. Even assuming that the existing efforts by Unilever and France on palm oil are successful, it will still be necessary for the plant to be planted and for it to grow until its productive age. This means that oil production will not be achieved till at least ten years from planting out. We also observed that application of genetic engineering in the seed industry and nitrogenous fixation are not likely to achieve concrete results in the foreseeable future.

Therefore, in the short term, the main impact on the agricultural sector will come from the diffusion and enhancement of technology already applied, e.g. rhizobium inoculant, and tissue culture in some plants like tubers and roots (e.g. cassava and potato), or simple plants (e.g. strawberries and ornamental plants). In all these cases, developing countries (though not all) are likely to benefit from the innovation provided that appropriate government policies are adopted, people are trained and mechanisms for the diffusion of technologies are introduced.

The situation with regard to pesticides seems to be a more complex one. In this case the interest of developing countries is in conflict with that of the multinationals which are active in the chemical sectors and are making efforts to secure markets. The Brazilian experience shows that biopesticide developments can be quite rapid, with favourable effects on the domestic agricultural sector.

The importance of the multinationals in biotechnology diffusion is also great in the food-processing sector. Evaluating the future impacts of biotechnology is very difficult for several reasons. The biotechnological innovations are taking place within an oligopolistic framework which maintains secrecy concerning their stage of development and characteristics. Such secrecy is particularly evident in the enzymatic industry which is concentrated (more than 60 per cent) in only two multinationals, viz. Novo industries and Gist Brocades. A similar situation exists with regard to the production of amino acids through fermentation. The production is dominated by three Japanese companies: Ajinomoto, Kyowahakko and Mitsui.

To sum up, biotechnology innovation and diffusion is an irreversible scientifically-based technological process. It can have negative or positive effects on the economies of developing countries, depending upon the economic and institutional mechanisms that each country will adopt for the application of these innovations.

At the international level, developed countries and multinational enterprises are making effort to apply biotechnology to solve their particular problems (agricultural surplus and dependence on imported raw materials) or to secure markets for their production (chemical sector). This is a fact that developing countries have to face.

NOTE

1. However, according to Professor Gerd Junne of the Universiteit van Amsterdam, an increase of isoglucose production in Canada would not have any negative impact on Canadian sugar imports because practically all isoglucose produced in Canada is sold in the United States. The isoglucose domestically produced cannot compete in the Canadian internal market because of the lower prices for sugar (world market level) as no protectionist measures such as those of the United States are imposed by Canada.

Part II
Advanced Plant
Biotechnologies

4 Advanced Plant Biotechnology in Mexico: a Hope for the Neglected?

Amarella Eastmond and Manuel L. Robert

INTRODUCTION

This chapter provides the setting for the three advanced plant biotechnology (APB) case studies in this volume. While the two subsequent chapters have the advantage of drawing conclusions based on *ex post* survey data, this chapter uses deductive reasoning to demonstrate how APBs can be deployed to deal with specific socioeconomic and technical problems. APB refers to the techniques developed over the past 20 years to manipulate plants in the laboratory including both cell biology (tissue culture) and molecular biology techniques, as has been elaborated in chapters 2 and 3 in the section on tissue culture.

This chapter focuses on four crops (coffee, henequen, coconuts and citrus fruits) totally bypassed by the Green Revolution, on which the poor depend for their survival in a harsh ecological environment – the south-east region of Mexico.[1] The approach followed consists of describing the problem and analysing how APB could provide the solutions. This includes an analysis of implications for employment, income distribution and growth. The crop case studies begin with coffee.

The chapter concludes with a discussion on Mexico's scientific and institutional capabilities in the generation of biotechnology.

COFFEE

As a source of agricultural employment, coffee is the single most important crop in the region. In some parts there is no alternative other than coffee growing for poor producers who increasingly need to supplement their subsistence agriculture with some form of cash income. It is estimated that about 400 000 people depend partially or totally for their livelihood on coffee production in Mexico as a whole (Nolasco, 1985). Of these, the largest and most impoverished group (270 000 persons) consists of agricultural day labourers, many of them from ethnic categories without access to land of their own and who work for coffee producers. Some of these labourers are also small coffee producers (accounting for 30 per cent of all producers) who are temporarily forced to sell their labour because of their precarious economic situation. The next largest group is made up of the 107 000 coffee producers. Of these the overwhelming majority (87 per cent) operate on a small scale, having an average of 1.5 hectares each.[2]

Not surprisingly, only the large, capital-rich producers can afford to make use of the Green Revolution technology and consequently they achieve almost double the yields of the small producers (16.6 sacks per hectare as against 9.3) (Nolasco, 1985). In addition, they have ready access to government and private credit and extension services, while many of the small ones (26 per cent of all producers) obtain no credit at all or, if they do, do so under very unfavourable conditions. Finally, the large producers have well-established commercial contacts both in Mexico and abroad which allow them to obtain better prices.

More than 90 per cent of the small plantations have been affected by the orange leaf rust disease which has spread over 100 000 hectares. The disease cannot be eradicated but can be contained by using cupric and systemic fungicides. However, such chemical control is beyond the reach of small producers owing to its cost and their lack of technical knowledge. Only producers with yields above 14 sacks per hectare can afford the cost, which, for each hectare treated, is roughly equivalent to the earnings from 2 tonnes of coffee berries (this includes the cost of labour but not that of the spraying equipment).

Without varieties resistant to orange leaf rust, all the coffee growers would experience diminished yields and increased production costs, but it is the small producers who stand to lose most because of the risk of seeing their whole crop wiped out for lack of chemical control. In the absence of a massive programme to provide free fungicides or credit for obtaining them, 75 per cent of the coffee producers might find themselves forced to abandon coffee and return to uncertain subsistence agriculture or to work on the big plantations as day labourers.

In these circumstances APB can make an important contribution to coffee production and the survival of the poor by producing, through micropropagation the resistant materials needed to control orange leaf rust. Mexico has, in fact, the capacity to develop and exploit the technology.

The application of APB to coffee would not only prevent small producers from losing their entire crop and livelihood but would also benefit the entire range of coffee producers. In terms of numbers and the impact on living conditions, however, the small growers stand to gain most if the work is carried out, as presently planned, under a government-sponsored programme to ensure that small producers have access (through credit and price control) to the technology. Not only would this protect the livelihood of more than 93 000 small producers and a large proportion of the 270 000 day labourers but, by substantially increasing yields, it could also raise wages and increase the number of rural jobs. It should be noted that 66 per cent of the coffee producers depend on unpaid family help (mostly women and children), particularly at times of peak labour demand.

Increased levels of production would generate additional employment through forward linkages to the coffee industry, which currently provides work for just over 4000 persons as well as in a variety of related activities such as transport, commerce and export. Employment would also be generated in laboratories and nurseries engaged in micropropagation.

HENEQUEN

The number of henequen[3] field workers in Yucatán, including *ejidatarios*[4] and small landowners, was estimated at 69 460 (in 1981), accounting for 60 per cent of the state's agricultural labour force. In addition, the number of people employed in the henequen textile industry was 12 500 in 1980, making a total of approximately 82 000 employed workers, or 22.3 per cent of the total labour force (López Huebe and de Fuentes, 1984).

The major constraint to the growth and development of this industry is a shortage of plantlets with which to replant the fields. In 1984 a government programme was drawn up for the planting of 63 000 new hectares by 1990. This target cannot now be reached before 1994 at the earliest because the planting of 50 million plantlets on 500 hectares of nurseries fell behind schedule.

To overcome this problem it is theoretically possible, using micropropagation techniques, to produce 1 million plants from a single mother plant compared with only 12 to 14 plants produced naturally by the same mother plant during its entire life cycle. In addition to the reproduction ratio, micropropagated plants demonstrate higher levels of productivity (they grow 70 per cent faster, produce more leaves and give off rhizomes when they are only one year old).

COCONUTS

The coconut industry provides work for about 50 000 people – mainly small producers, *ejidatarios* and agricultural day labourers – in Mexico, including all five states in the south-east. Because of the low level of technology used, inadequate application of inputs (irrigation and fertilisers) and poor management, yields are low.

Coconut palms in Mexico are now threatened by the spread of 'lethal yellowing'. This disease, which kills trees within five months from the moment the first symptoms appear, is caused by a mycoplasma-like organism that develops in the sap of the infected trees. The pathogen is transmitted from diseased to healthy individuals by a small beetle (*Myndus crudus*) which feeds on the sap of the trees.

There is no cure for lethal yellowing; the symptoms can be delayed by applications of oxytetracyclin but the trees are doomed as soon as the antibiotic is withdrawn. To some extent, its spread can be halted by spraying insecticides. However, the beetle lives in the surrounding grasses and it is impracticable to spray a large enough area for this to be effective. In recent years, the disease appears to have been advancing at an average pace of 25 kilometres per year.

In vitro tissue culture offers the only possibility for the asexual propagation of coconut palms; this would not only provide disease-free planting material but also make available a larger number of plantlets at greater speed.

Through the application of APB in controlling the lethal yellowing disease, coconut cultivation could serve as a major supplementary source of employment (on average 45 work-days per hectare annually) for the poor to combine with other occupations (most commonly fishing and salt production in Yucatán).

CITRUS FRUIT

Applications of APB to the citrus fruit crop would cut the costs of chemical means of disease and pest control for farmers in the south-east.[5] Its use to produce virus-free citrus material would substantially reduce crop losses (currently 50 per cent) due to pests and diseases, at a fairly low cost. Since APB would initially focus on disease-free material, it would not displace labour on the labour-intensive operations of weeding, pruning and irrigation.[6] Moreover, any loss in employment in disease and pest control would be more than compensated by the greater labour input required for the harvesting and transport of the larger output.

Small producers (with less than 3 hectares) account for nearly all (98 per cent) of the 5246 farmers engaged in citrus cultivation in the Puuc region of Yucatán (Table 4.1). By reducing crop losses, the application of APB to citrus fruits would enhance their earning capacities. It would also increase the demand for day labourers – who, at over 3000, account for 36 per cent of the citrus industry's labour force – and thereby help to boost their wages.

A number of jobs are created indirectly through the forward linkages to the juice processing plant (Table 4.1). Twenty-one industrial workers are employed by the Union of Orange Growers in the concentrated juice plant in the vicinity, while 11 men are full-time lorry drivers and 34 are market vendors. Observation on several market days suggests that there are frequently as many as 30 people carrying produce for clients. Thus orange growing provides employment for some 8400 people in all, or about 37 per cent of the region's labour force – though these figures no doubt underestimate the real numbers. APB could help to provide employment for farm workers made idle (underemployment rates of 75–90 per cent) because of the seasonality of citrus fruit production by prolonging the season of ripe oranges, which could be achieved by planting more varieties. At present the juice processing plant operates

TABLE 4.1   *Citrus fruit production and employment
in the Puuc region of Yucatán, Mexico, 1987*

Type of employment	Number of jobs	% of total jobs
Total producers	5246	62.4
Of which:		
Small producers (less than 3 ha)	5123	(97.7)
Medium producers (3–10 ha)	100	(1.9)
Large producers (over 10 ha)	23	(0.4)
Farm overseers	12	0.1
Agricultural day labourers	3052	36.3
Employees in juice plant	21	0.2
Lorry drivers	11	0.1
Market vendors	34	0.4
Carriers	30	0.4
Total	8406	100.0

SOURCE: Adapted from SARH, Centre for Rural Development Support No. 2, unpublished citrus production data 1987, and producer interviews in Oxjutzcab, 1987.

only six months a year (September to February), when oranges of the right quality are available.

Indirect employment would also be created in the laboratory and green-houses that would need to be installed and that would probably be able to produce 500 000 plants a year (the amount currently sold in the region). Growers now buy their scions grafted on sour orange (at US$0.40–0.50 per plant) but are given no guarantee of the scions' quality. Although it is difficult to estimate the costs of virus-free citrus material, more than half the growers said they should be able to afford it and would be interested in buying the material if it paid off economically. APB therefore offers the possibility of increasing growers' incomes, providing more work for day labourers, reducing the idle capacity of the juice processing plant (thereby increasing the number of industrial jobs) and stimulating the local economy. The juice processing plant has already generated demand for field labour and real incomes have risen significantly. Informal sector jobs related to agriculture have also been created by the plant. In fact, the multiplier effect of the plant has been such that a reversal in the relative incomes of agricultural producers and petty officials has occurred (the former can now earn two or three times the annual salary of the latter).

These relative changes in incomes and soaring land prices have resulted in some social inequality. A simplified classification based on occupational

TABLE 4.2   *Changes in class structure in the citrus fruit zone of Mexico,*
1970 and 1980

Occupation	Class	*Percentage of total occupations*	
		1970	*1980*
Public officials, managers, administrators, large landowners	Upper	1.4	0.3
Professionals and technical workers, administrative personnel, traders	Middle	16.2	4.7
Personal service workers, transport workers, small agricultural producers, cattle-men, lumbermen, fishermen, hunters, agricultural day labourers, non-agricultural workers, machine operators	Lower	82.4	78.0
Unspecified and unemployed	Lower	–	17.0
Total		100.0	100.0

SOURCES: 1970: Webber, 1980, for Yucatán; 1980: Adapted from Secretara de Agricultura y Recursos Hidráulicos, 1984.

status shows a markedly more skewed distribution in 1980 than in 1970 of the population among the various occupational classes (Table 4.2). This trend of increasing social differentiation may be further aggravated, although labour's high factor share (about 80 per cent of total costs is accounted for by labour costs) persists.

Mexico's indigeneous institutional and scientific capabilities

With very few exceptions (amongst which Brazil and India are outstanding), most LDCs suffer from some or all of the following obstacles to developing APB: (a) lack of a well-defined National Programme on biotechnology; (b) a shortage of highly-qualified researchers; (c) weak links between basic and applied research; (d) limited economic resources; (e) small private participation in the field; and (f) limited economic feasibility of the technologies.

National programme on biotechnology
Mexico's potential is limited by the lack of a well-defined national policy to develop biotechnologies. It should be noted that biotechnology is recognised by the authorities as a potentially important area for the country's commercial and economic development and has been assigned top priority by the National Council of Science and Technology in its programmes in accordance with the national programme on science and technology (Gobierno Constitucional, 1984a). However, in spite of the fact that some analytical studies have been carried out (Quintero, 1985 and CEPAL, 1988), no specific areas or lines of research have been defined as national priorities. Even where certain problems have been identified, action is slow to follow or is not taken at all. As an example, we can cite the case of *lethal yellowing*. The disease was unmistakably diagnosed in Mexico in 1982 and the need was clearly established in a meeting in Merida in 1984 to develop PTC techniques for the micropropagation of coconut palms and to carry out basic research to develop diagnostic methods. However, it was not until 1988, mainly through individual initiative, that some work on the topic began.

Without a general framework and the identification of priorities, the allocation of resources and the selection of applied lines of research has been based on the individual interests of scientists.

Qualified human resources
In the case of Mexico the lack of highly-qualified human resources is a limiting factor but not a crippling one. In comparison with the United

States, Europe and Japan the Mexican research capability is very small. Nevertheless there are already some research centres that are capable of adopting or developing plant biotechnologies and several M.Sc. and Ph.D. programmes that should enable the country to increase its qualified human resources fairly quickly. On the other hand, there is a growing awareness and interest in APB – the subject is taught at B.Sc. level in most biological sciences and agricultural curricula.

Links between basic and applied research
Many scientists in Mexico consider that the pursuit for knowledge should be the one and only objective of scientific research. They look down on those who get their hands dirty searching for solutions to practical problems. Biochemists and molecular biologists consider that agronomists are closer to farmers than to scientists, while the latter accuse the former of locking themselves in 'ivory towers'. Until recently both groups could survive without much interaction but with the development of APB, this situation has changed. There is an urgent need for closer collaboration if APB is to be successfully applied. As yet it has proved very difficult to bridge the gap between the two groups.

Limited economic resources
It is unrealistic to expect that much of the R & D done in developed countries will be of direct benefit to the agricultural challenges of Mexico and particularly those of the south-east. Firstly, because developed countries are concerned primarily with solving their own problems and maximising profits, and secondly, because the agricultural characteristics of the south-east make it an unattractive market. As L.V. Mayer (1988), Deputy Assistant Secretary for economics of the US Department of Agriculture, succinctly told the Agbiotech 88 meeting in Washington: 'The low incomes of the hungry prevent them from becoming viable commercial customers'. R & D from developed countries will therefore largely reach LDCs through two channels: (a) incidentally, as a by-product of research done for other purposes, and (b) if it is directed through multinationals which have a stake in the region. In both cases, patents will ensure that LDCs pay dearly for the new technology. At present, Mexico does not recognise plant patents but this is due to change in the near future. Therefore, the implications are that if it wants to make use of APB, Mexico will have to do most of the research itself, which will be a very expensive undertaking.

Until now most APB research projects in Mexico have been financed by the grant systems of CONACyT and COSNET and by SARH. However, during times of financial crisis there is little hope of sufficient public funds being channeled to research institutions to ensure an adequate transformation of ideas into reality. The new policy of making the government research centres economically as self-sufficient as possible clearly manifests the government's intention of reducing its support for basic research.

The role of the private sector
For a biotechnology to have an economic impact it is essential that there is a user that not only needs it but that has the financial possibility to adopt it. Many research projects on APB in Mexico are undertaken on the researcher's initiative in the hope that the results will convince someone later on. Consequently, many of them have to be abandoned due to lack of financial support or because of the impossibility of transferring a product to the field for evaluation. If anyone in Mexico can make use of APB it will be the private sector. Not only does it have the economic capacity to fund the necessary research, but as an early adopter, it will be able to capture and dominate the markets. Furthermore, entrepreneurs are receiving the lion's share of government backing. Firstly, they obtain through tax deduction incentives to finance research. Secondly, the government is now encouraging its research centres to turn to private industry for contracts and is therefore putting at the latter's disposal a very large and expensive research capacity. Finally, the shared-risk schemes available through CONACyT and Nacional Financiera (NAFINSA) to finance research and development, which are nearly at the production stage, remove some of the economic risk from the projects for the client.

Nevertheless, private capital will only be put forward for biotechnology ventures where there is a clear market for the end product. Looking at agricultural producers in Mexico, only a very small percentage (less than 2 per cent) can be classified as large producers (CEPAL, 1982) who would have access to sufficient resources to invest in APB. They would only take the decision to invest if the price they will get for their products justifies the additional input costs. The guarantee prices of maize, beans, sugar and many other crops are too low to stimulate further investment. If left to market forces, then it is difficult to see how APB will be applied to the crops that could positively affect the vast majority of producers and consumers. There is also great scepticism among the private sector that the new techniques will live up to their promise. Producers are very reluctant to invest in something that has not yet proved to be of superior quality in the field, especially since the returns are unlikely to accrue for several years. In the cases we have analysed here neither the *henequen* nor the coffee plants that are being tested in plantations have proved their merit in a conclusive manner. In the same way, the suggestion that virus-free orange trees derived from tissue culture are going to increase yield is nothing but a theoretical possibility to the producers of the southeast.

With the exception of the cut-flower industry in which a great deal of foreign interests exists, only the tequila industry and a couple of other companies have taken the initiative. They are financing APB projects under confidential agreements with public research centres.

It is fundamental that whoever finances APB, whether the government, private industry or producer associations, they should be conscious of the limitations and difficulties involved and be prepared to sustain multi-disciplinary programmes for many years beyond the boundaries of the six-year government periods. It is relevant to point out that the tissue culture work on oil palm at IRHO started in 1970 (Sasson, 1988) and the definitive evaluations on the performance of the *in vitro* produced plants are not expected until 1990, 20 years later. It should also be borne in mind that there are no permanent solutions to disease and pest resistance because of the continuous evolution of pathogens and predators. It is therefore essential to carry out continuous and long-term APB programmes in an attempt to keep pace with natural changes.

Economic feasibility
The potential of a specific plant tissue culture technology is usually based on positive results reported in research papers. However, in many cases the plants produced have never left the laboratory, probably not even the test tube. As commercial micropropagators know only too well, this information is of no practical use since it pays little or no attention to many very important aspects such as multiplication and survival rates, genetic stability and trueness to type of the products. Also, it completely disregards the costs of producing thousands of the new plants with the method.

Estimating costs is very difficult because of the many factors involved. For instance, labour, which accounts for 60–70 per cent of the total cost, is particularly variable, being dependent on skill and efficiency. High inflation rates in Mexico add to the difficulties (during 1987 the inflation rate reached 150 per cent). Micropropagated plants are always considerably more expensive than traditionally-propagated ones. This is understandable since micropropagation techniques are labour intensive and one has to add the expenses of weaning in mist or green houses and time in the nurseries to those of the laboratory. It is therefore not surprising that the private sector has been so reluctant to invest in something whose merits have not been proven in the field and the profits from which will not start to flow for several years.

Jones (1982) has calculated that each cloned oil palm plantlet costs five times as much as a seed propagated one but that the extra cost would be rapidly recovered if there were an increase of 30 per cent in oil yield and that profits can be made after the fifth year of harvest. He also estimates that to be cost effective a micropropagation unit must produce in excess of a million plantlets per year.

Although manual labour is relatively cheap in Mexico (at least 10 times cheaper than in the United States) it is imperative to reduce costs before tissue culture can become more widely applied to low-value crops. The formation of somatic embryos from single cells grown in fermenting tanks is at present the most efficient system known to produce large numbers of plants in short times with relatively little labour. In addition, somatic embryos can be encapsulated in alginates to form artificial seeds which will further reduce costs, making handling easier (Fujii *et al.*, 1987). None the less, the technique cannot be applied yet to many species and poses the risk of genetic instability. It might be possible to apply it more widely as our knowledge of the mechanisms that control differentiation and gene expression increases.

Automation of certain tissue culture processes is also progressing (Levin *et al.*, 1988) but as yet it is unknown whether they will be able to compete with Mexico's cheap labour.

Capabilities with case study crops

The four case study crops we have analysed in this chapter illustrate some of the general points raised about Mexico's capabilities.

The advance in some fields and in certain crops cannot be taken as a basis for predictions in other species and even results reported in the scientific press may still be far away from application. For example, it was thought that micropropagation work on coconut palms would follow in the footsteps of the oil palm work done by Unilever in the United Kingdom and IRHO in France, which is already commercialised in several parts of the world (Corley, 1983; Sasson, 1988). An optimistic prediction by Branton and Blake (1983), when they first produced a clonal coconut plant, has permeated to Sasson (1988) and Tudge (1988) who report that, from the work at Wye and IRHO, coconut micropropagation is ready for large-scale commercialisation. Tudge (1988) also reports that at Hindustan – Lever (Bombay, India) coconut micropropagation is 'on the point of commercial success'. However, at least to our knowledge, there is still no available technique for the efficient micropropagation of coconuts. Furthermore, the abnormalities found in Unilever's oil palm plantations in Malaysia indicate that the cloning methods used for oil and coconut palms with long periods *in vitro* are not safe.

The case of citrus illustrates a totally different example of how a proven biotechnology developed in other countries, in this case the United States and Spain, could be exploited by LDCs in the short term. The *in vitro* quarantine methods and release of disease-free budwood for commercial propagation as developed by the citrus variety improvement programme of Spain, can be transferred to any country for its own benefit. Navarro has made very detailed recommendations as to how the quarantine programmes should be carried out and the *Instituto Valenciano de Investigaciones Agrarias* trains scientists from other countries in order to help implement them. However, growers in Mexico are not easily convinced of the benefits of these programmes since it is difficult to quantify the financial gains due to reduced losses. And at any rate those gains are long term.

The agave is an example of a very successful methodology developed in Mexico for a very specific purpose. However, it illustrates once more that the factors determining whether a biotechnology will be applied or not do not rest on its potential or on the need for it, but on who can pay for it. It is doubtful that the production of *henequen* will receive the potential benefit of an increased yield (that would have such a positive impact on the thousands of *henequen* workers in Yucatan) because of the low world price and the present *henequen* production system. Nevertheless, the privately-owned *tequila* industry is financing agave production in the laboratory and carrying out field tests. The industry can afford the expense since it is a high-value export product which makes it easy to recover the initial investment in a relatively short time.

The case of coffee could be one example of APB reaching the producers since the government institution, INMECAFE, has already taken the first steps by financing the research and is now testing the new resistant plants in the field. It is yet to be seen how good their performance is and at what cost they will be distributed. It should be mentioned that there is opposition to the use of APB techniques for the control of orange leaf rust amongst agronomists who think that the disease can be more efficiently controlled with traditional methods.

CONCLUDING REMARKS

It appears that APB will only be adopted to a limited extent in Mexico in the short to medium terms and largely within the private sphere. It follows that the majority of its direct benefits will also be restricted to certain high-value crops and to those who produce and consume them. Therefore, poor consumers cannot expect to benefit from a general reduction in food prices, as occurred in the GR, at least for a while. Indeed, the relatively greater profits from high-value crops will continue to encourage farmers to abandon staple crops, making food scarcities more probable.

While the adoption of APB in specific crops such as bananas, coffee and coconut is unlikely to reduce employment in the south-east, it is doubtful whether it would directly create many new jobs for agricultural workers and producers despite the optimism emerging from the simulation approach. Some privileged areas, such as the Puuc region, might be able to intensify production, but in comparison with the magnitude of the problem of rural under – and unemployment it is unrealistic to expect APB as a panacea. Moreover, the majority of new jobs would be the ones for day workers at a minimum wage. In the short term, the creation of independent jobs depend more on the provision of irrigation and communications infrastructure by the government than on new agricultural technology.

Job loss from APB-facilitated mechanisation, on the other hand, is not an immediate threat in the south-east because of the adverse ecological conditions. Mechanisation is only likely to intensify at the food-processing stage and only on a small scale, due to the relative cheapness of labour and the lack of capital for long-term substitution.

The adoption of varieties which no longer require pesticide protection or fertilisers would slightly diminish labour demand but the benefits to human health and the environment would more than compensate for this. On the positive side, a few technical jobs would be created to run the APB laboratories.

In summary, it would appear that APB in itself is not one of the keys to socially and environmentally-sustainable development in the south-east of Mexico although it could be a powerful secondary tool. Ultimately, it is the underlying structure of Mexican agriculture and the rural development policies adopted that will determine the future agricultural growth and the distribution of the benefits. So long as present imbalances regarding access to credit, irrigation and the commercialisation of products are not corrected, APB adoption will continue the trend of favouring commercial agriculture and high value crops, while its potential to relieve poor consumers and producers and diminish technological and food dependence will largely remain dormant.

NOTES

1. Comprising the states of Campeche, Chiapas, Quintana Roo, Tabasco and Yucatán, this region consists mainly of humid and subhumid tropical lowlands a large proportion of which is unsuitable for farming; Yucatán has very little good arable land and three out of the five states are in part badly waterlogged. While the rest of Mexico has an average of 27.7 per cent of irrigated farm land, only 1.6 per cent of the south-east is irrigated. It is believed that the underemployment rate (people earning less than the minimum wage) ranges between 75 and 90 per cent for those engaged in agriculture.
2. This group owns only 35 per cent of the land planted to coffee and produces 18 per cent of the harvest. In contrast, 2.5 per cent of the producers with plantations of 100 or more hectares own 29 per cent of the coffee land and produce 28 per cent.
3. A natural fibre used to produce rope and twine.
4. *Ejidatarios* are peasants cultivating communal lands who are granted lifetime usufruct rights by the Government under the land reform programme introduced after the 1917 Revolution. The system was designed as the basis for an equitable distribution of land to the entire rural population.
5. Data were obtained from estimates by SARH and direct interviews of growers in the juice processing plant.
6. The highly labour-intensive character of citrus fruit production (between 78 and 82 per cent of the production costs is accounted for by labour) would be preserved (as would be labour's factor share).

5 Biotechnology and Farm Size in Kenya

L. P. Mureithi and B. F. Makau

INTRODUCTION

This chapter is concerned with the biotechnology applications to two important crops, potato and tea in Kenya. Potato is an important source of food for the Kenyan poor, one third of which is grown by the small holders. Kenya is the world's fourth largest tea exporter. Tea is also the second largest foreign exchange earner for Kenya. More than one in every ten small-scale farmers grow tea on their farms. The crop is thus indicated as being of great importance to the nation in terms of the generation of income, foreign exchange and employment.

PURPOSE

The main APB applications in Kenya are meristem (tissue) cultures leading to clonal multiplication of potato and tea crops. The purpose of this chapter is to examine the impact of these applications on (a) labour absorption; (b) backward linkages to agricultural input suppliers; (c) efficiency, land and labour productivity; (d) profitability; (e) the rural labour market and labour's factor share, and (f) income distribution. The chapter also examines the relationship between farm size and most of the above variables in particular land productivity.

The chapter begins with the study of potato followed by that of tea. The section on each begins with an elaboration on the sample survey design used to generate the data.

POTATOES

Technological trends

Excluding the traditional plant breeding, tissue culture work on potatoes is undertaken at the National Plant Quarantine Station (NPQS) of the Kenya

Agricultural Research Institute at Muguga. Its function is to screen and multiply imported material such as those from the International Potato Centre (CIP); to clean local material for fresh clones; to undertake limited multiplication of material to meet Kenya's needs; and to preserve the germplasm received.

The current method commonly used at the station utilises the apex of the plant for tissue culture. Material may be brought in the form of tubers or seeds or seedlings and/or meristem tissues as is the case with CIP material. The material is germinated (from seeds, tubers or meristem tissue) and small pieces of apex tissues are removed and grown in special growth media under controlled conditions. At the same time, the tissues of the germinated plant are tested for diseases – the most common method for testing the viruses at the station being the ELISA serological method. The plantlets developed from the apex tissue are tested further. If found to be free of viruses, multiplication is undertaken on a larger scale through the same method and the resulting material is supplied to the National Potato Research Station (NPRS) at Tigoni for clonal development.

The NPRS was acquired as a farm (formerly called Mthanga farm) in 1967. In 1972 it was designated a research station. It has developed as a disease-free centre for raising and screening seedlings for blight resistance and for the multiplication of healthy stocks of established varieties and new promising cultivars. In more recent years, it has incorporated a research programme in breeding, pests, agronomy, processing and socio-economics. However, to date maintenance of clonal breeding and the provision of enough seed for foundation stock remains the largest programme at the station. This work is undertaken both at the main station at Tigoni and at sub-stations in South Kinangop, Meru and Molo. This makes it possible to obtain information from different ecological zones and, at the same time, ensure a supply of adequate foundation stock.

Foundation stock is multiplied by large-scale growers and once the seeds have qualified after inspection they are called basic seed. These are sold to farmers who wish to produce certified seed. Table 5.1 gives the hectareage of certified potato seeds grown since 1979. Indirect credit to large potato growers and the occasional certified seed outgrowers is in the form of certified seed, chemicals and gunny bags, the cost of which is subtracted from the sale proceeds. Grading, storage and marketing is done free of charge for the farmers participating in the ADC outgrowers' schemes.

The ADC has good storage facilities where up to 2250 tonnes (45 000 bags) can be stored for up to six months at about 4°C. This cold storage started operating in 1985. The Agricultural Development Corporation (ADC) markets the potatoes through the Kenya Grain Growers'

TABLE 5.1 *Production of certified potato seeds*

Year	Hectares approved
1979/80	120
1980/81	467
1981/82	414
1982/83	362
1983/84	280
1984/85	364
1985/86	444

SOURCE: Records at the National Seed Quality Control Centre.

Co-operative Union (KGGCU) which has agricultural inputs and outlets for products throughout the country. However, the KGGCU does not itself buy to sell. It only stocks the potatoes for the ADC and the latter has ultimate responsibility for the stock, whether or not they sell.

The survey design

We purposely selected 33 farms growing potatoes in 1986, bearing in mind the potato zones of the small-scale and large-scale farming areas. For the former, we chose Kabaru on the slopes of Mount Kenya and Othaya on the slopes of the Nyandarua ranges, while for the latter, Molo and Mau Summit in the Rift Valley highlands proved ideal. The majority of the potato farms are small, being under 2 hectares. Anything in excess of 10 hectares is large and so between 2 and 10 hectares was classified as medium size. Data were gathered through direct interviews using a structured questionnaire.

The diffusion of innovation

The extent of the adoption of clonal certified seed potatoes can be gauged by the fact that 27 per cent of all the farms in our sample had certified potato fields as pure stands while 36 per cent of them had mixed (certified and non-certified) stands. The crude adoption rate, that is, the proportion of farmers adopting clonal certified seeds, is positively associated with farm size. Taking the mixed fields, the adoption rate was 25 per cent for the small-scale farms, 46 per cent for the medium-sized farms and 50 per cent for the large-scale farms (Table 5.2). The propensity to adopt certified seed is not systematically related to education, since the average length of education is 7.3 years for small-scale farmers, 4 years for medium-scale farmers and 6.5 years for large-scale farmers.

TABLE 5.2   *Adoption of innovations: potatoes in Kenya*

Farm Size	Percentage of farms with			Percentage of cropped areas with		
	Certified crop	Non-certified crop	Mixed field	Certified crop	Non-certified crop	Mixed field
Small	38	38	25	21	48	28
Medium	15	38	46	17	46	36
Large	25	25	50	37	12	51
All	27	36	36	28	28	42

The intensity of adoption, that is, the proportion of potato area brought under clonal certified seeds, increases from 21 per cent in the small farms to 37 per cent in the large farms. This is possibly due to the fact that the Agricultural Development Corporation's potato project deals quite aggressively with the large-scale farmers, some of whom are used for purposes of field multiplication of certified seeds. It is instructive, however, that the adoption intensity declines with farm size – it being 17 per cent in the medium-scale farms – when large farms are excluded. We conclude, therefore, that the intensity of adoption is not positively associated with farm size, other things being equal.

Income generation and distribution

As a business, a farm generates income in cash and in kind. Total output – taken to include the market value of potatoes sold, fed to livestock or used for own consumption – amounted to Ksh. 37 269 per hectare of certified potatoes in small farms while the same area yielded Kshs. 20 641 if planted with non-certified seeds. For large farms, the yield was Ksh. 29 273 for certified and Ksh. 13 265 for non-certified potatoes.

The following should be noted. Firstly, the gross output per unit area is higher for certified potatoes than for the non-certified variety, whichever farm size category is considered. This is to be expected since the certified varieties of potatoes yield a larger physical output per hectare than the non-certified varieties, in our case 36 721 kilograms versus 14 406 kilograms. Secondly, gross output per hectare declines as farm size increases. This collaborates the 'large body of evidence that suggests that small farmers make more efficient use of available land than large farmers' (Hayami and Ruttan, 1985).

In Table 5.3, we present statistics on production, inputs, gross margins and value added. Intermediate inputs are seed potatoes, fertilisers and other

TABLE 5.3 *Output, input and surplus (shillings per hectare)*

Indicator	Certified potatoes				Non-certified potatoes			
	Size of farm				Size of farm			
	Small	Medium	Large	All farms	Small	Medium	Large	All farms
Gross output (1)	37269	25580	29273	33210	20641	21744	13265	16362
Intermediate input (2)	4171	5004	1774	3553	3101	2433	2621	3008
Equipment usage (3)	289	484	1135	867	157	385	556	426
Labour costs (4)	3621	9403	7246	7974	2925	3756	2164	3012
Operating costs (2+3+4) = (5)	8081	14891	10155	12394	6183	6574	5341	6446
Operating surplus (profit) (1–5)	29188	10689	19118	20816	14458	15170	7924	9916
Value added (1–2)	33098	20576	27499	29657	17540	19309	10644	13354

SOURCE: Survey data.

chemicals such as insecticides, pesticides, weedicides and fungicides. We also include in this category transport for both the intermediate material inputs and potatoes. Intermediate inputs, therefore, largely represent the extent to which the potato industry patronises other sectors and to that extent stimulates them to produce further output, generate more employment, etc. Taken as a proxy for backward linkage, intermediate inputs per hectare are higher for certified than for non-certified potatoes.

Capital equipment in potato production includes forked and plain hoes, pangas (machetes), sprayers, *debes* (tins) and an occasional tractor. We assumed that the useful life of such equipment is an average of ten years. On this basis, 10 per cent of the value of the equipment was taken to be consumed in potato production every year, that is, the flow of capital services. As expected, capital usage is less in smaller-scale farms than in larger ones which tend to be more mechanised. As Table 5.4 reveals, certified potatoes exert greater demand for capital usage than non-certified potatoes.

Labour costs were calculated for hired and family labour, the latter being valued at the going wage rate for hiring farm workers, in other words the valuation is on the basis of opportunity cost. In general, it takes more work-hours to produce clonal potatoes, particularly due to the cultural practice of ridging before planting certified seeds – a practice that is not universally observed in non-certified potato cultivation.

The total costs of operation per area of cultivated land are higher with innovation than without, just about double. It is true also that the profits in such enterprises are in roughly the same order of magnitude. Yet, the fact that one has to incur the cost prior to drawing the benefits might act as a budgetary (financial) constraint to innovation. This could explain the somewhat reluctant manner with which small farmers innovate, as reflected in Table 5.3 by the relatively low land area devoted to clonal certified potatoes. We therefore feel that there should be in position a credit scheme to enable farmers to adopt proven biotechnical innovations. Such a scheme could use improved potatoes as an anchor crop and would cover, not just the acquisition of seeds and intermediate materials, but also the cost of labour since 'cash flow constraints especially hinder fertiliser use and labour use' (IFPRI, 1986).

The average wage rate ranges from Ksh. 19 per work-day in medium-sized non-certified potato farms to Ksh. 29 in large-scale certified potato farms. The statutory daily minimum wages range from Ksh. 17 for rural areas to Ksh. 31 for municipalities. Hence, labourers deployed on potato production are no worse off, whether the labour demand is seasonal or casual. A further indication as to whether one can make money in potato activity is gauged by the farm revenue and the owner's operating profits. With gross revenue at an average of Ksh. 33 210 for a hectare of certified potatoes, this compares favourably with other possible agricultural activities such as coffee at Ksh. 29 780 and vegetables at Ksh. 18 260 per hectare in 1984 (Kenya (a), 1986). If the farm operator opted to be employed in a modern sector job, the average wage earning would have been around Ksh. 20 958 in 1985 (Kenya (b), 1986) while foregoing Ksh. 20 816 as clear profit for growing one hectare of improved potatoes. Clearly then, the growing of improved potatoes plays a crucial role in the alleviation of poverty and warding off rural-to-urban migration – all salutary impacts on the employment problem. Of course, increased supply of potatoes, unless matched by increased consumption, may not lead to increases on gross farmers' revenues if prices decline.

Value added is gross output net of intermediate inputs and gives the earnings by the various factors of production – land, labour, capital and entrepreneurship. It is part and parcel of national product, i.e. national income avoiding double counting. From Table 5.3, it is clear that the contribution to national income is higher with new biotechnology than with the traditional technology.

Aside from the absolute level of incomes, the distribution of income is important. To assess this, we took the ratio of the share of gross income by the richest 30 per cent of the farmers to the share by the poorest 70 per cent of the farmers as the measure of concentration. This turned out to be eight for clonal certified potatoes and four for non-certified potatoes, as is shown in Table 5.4. It would appear that the 'Green Revolution' effect is being felt, whereby the introduction of modern varieties of crops acts 'as a source of inequity in income distribution' (Hayami and Ruttan, 1985).

Underlying the unequal distribution of income is uneven distribution of land. The smallest 70 per cent of the farms accounted for 10 per cent of the land area so that the coefficient of concentration of land ownership among

TABLE 5.4 *Distribution of incomes and land*

	Share of incomes		
Percentage of sample	Certified	Non-certified	Percentage of land area
Poorest/smallest (40)	3	7	3
Middle (30)	8	15	7
Richest/largest (30)	89	78	90
Concentration ratio	8	4	9

SOURCE: Survey data.

potato farmers is estimated at nine; that is, the sharing of this productive asset is even more unequal than its fruits. Perhaps this pushes the smaller farms to be more intensively cultivated to excel in productivity per land area in an attempt to catch up with their larger counterparts. While potato farmers may not be typical of farmers and our sample may be too small for very firm conclusions to be drawn, our data are nevertheless suggestive of highly skewed distribution of incomes and land such that 'the gains in production' arising from modern-variety technology 'have been offset by losses in equity' (Hayami and Ruttan, 1985).

Employment potential

In this section, we shall examine the employment implications of the introduction of improved varieties of potatoes. In empirical economic literature, the share of wages in value added has been shown to be an approximate measure of labour intensity (Lary, 1968). On this score, certified potatoes as a group, at 27 per cent, fare better than the traditional varieties at 23 per cent and our results support the hypothesis by Bhatty that 'biological technology, in general, is not known to be labour substituting in itself' (Bhatty, 1978).

An impression on the prevailing average capital-labour ratio is obtained by glancing at Table 5.5, where capital use per work-day is given. This reveals the characteristic positive relationship between capital intensity and farm size, and therefore the decreasing relative ability to create jobs as farms expand. This phenomenon in mechanical technology applies equally in certified potato operation as in non-certified fields, so that, on the average, capital-labour ratio is the same. To the question posed by Hayami and Ruttan (1985) as to whether modern-variety technology promotes mechanisation, our answer would be: no.

For a typical farmer, land is a fixed factor of production in the short run, and highly inelastic in supply even in the long run. For Kenya as a whole, it is fixed; it is therefore scarce and exerts an effective constraint. It should be of interest to see how use of improved seed varieties enables it to generate employment. Measured in terms of labour-land ratio – work-days of labour employed per hectare – certified potato farms are typically better at employment creation than their less innovative counterparts.

In its enhanced labour absorption, does improved potato cultivation operate at the expense of productivity? No. Because physical labour productivity (kilograms per work-day) is actually 24 per cent higher in certified potatoes taken together than in non-certified potatoes. Again, 'a rise in the proportion of value added (to gross output) reflects a

TABLE 5.5   *Labour intensity and factor productivity in potato farming*

Type of crop	Certified potatoes				Non-certified potatoes			
Size of farm	Small	Medium	Large	All farms	Small	Medium	Large	All farms
Value added to gross output (%)	89	80	94	89	85	89	80	82
Wages in value added (%)	11	46	26	27	17	19	20	23
Work-days per hectare (No.)	181	446	253	301	143	200	97	144
Capital use per work-days (Ksh.)	2	1	5	3	1	2	6	3
Labour productivity (kg/work-day)	124	27	170	124	61	67	193	100

SOURCE: Survey data.

lower input of materials for producing the same output. This is one of the principal objectives of technological improvements' (Bhatty, p. 78). Relative efficiency as measured by this criterion puts certified clonal potatoes ahead of the unimproved crops.

We conclude that the introduction of improved varieties of potatoes leads to more labour absorption on the farms and that this happens simultaneously with gains in factor productivity and relative efficiency. Although the increased income from such innovation is accompanied by worsening income distribution, the benefits derived from the multiplier effect on income and employment in the economy could offset the adverse socioeconomic effects of this increase in income inequality.

Biotechnology applications in tea

The major thrust in biotechnological applications in tea in Kenya is in the area of clonal development and vegetative propagation. The early propagation of tea was through the use of unimproved seed as a source of planting material. However, the use of seed resulted in high variability in various characteristics among individual plants. The programme of clonal development was started at the Tea Research Foundation (TRF) based in Kericho and in some individual plantations to reduce this variability. This effort has resulted in various clones which have uniform or less variable characteristics and which give superior yields and quality.

As a result, large plantations which were wholly planted with seedling material began a steady programme of replacing and filling with clonal material. To date, about 10 per cent of the large plantations are now planted with clonal materials and any new additions or replacements are always planted with clonal materials. The small-scale farms, all of which started after the initiation of the clonal development programme, are all planted with clonal materials. The tremendous expansion of tea production in Kenya in acreage and yields – especially small-scale production – owes its performance, at least in part, to clonal development and rapid vegetative development work.

A clone takes 12 years to select, develop and test. After this, elite cuttings are used for rapid vegetative propagation methods and this supply becomes generally available to farmers.

The principal tea-growing areas of Kenya lie between 1500 and 2200 metres above sea level, in areas enjoying between 1200 and 1700 mm. of rainfall per year, with an all-year cool, almost temperate climate. Hence the range of environmental conditions under which tea is grown is quite narrow, as the tea plant itself is sensitive to changes in environmental conditions.

The data for tea

A structured questionnaire was administered to 39 tea farms. These included *large-scale* estates in the Kericho district of the Rift Valley province and in the Limuru area of Kiambu district in the Central Province, where tea was introduced prior to Kenya's independence in 1963; and also small-scale farmers from Othaya in the Nyeri district and Githunguri in the Kiambu district who began growing tea under the patronage of the Kenya Tea Development Authority (KTDA) set up in 1964 for the purpose of promoting tea growing in erstwhile non-scheduled areas.

On the basis of information gleaned from various bodies in the tea industry we grouped tea farms into three classes: small – up to 3 hectares in area; large – in excess of 20 hectares; and medium – in between the above two. Detailed information was sought for the year 1986. Bearing in mind that human recall on data – particularly for small-scale farmers who generally keep no written records – would tend to diminish with time, we focused mostly on the previous year.

The KTDA introduced improved tea varieties right from the beginning so that – in general – smaller tea farm operators grow relatively more recent clone vintage than the larger estates, which – again in general – grow the more traditionally bred tea varieties. However, although tea is a perennial crop which could sustain production for decades on end – and which it would be agonising therefore for the farmer to cut down once planted – some plantations have found it wise to replace their seedling-based crops with clonal-based crops. They also invariably plant any newly expanded tea areas with improved varieties such that the proportion of area grown with improved clonal tea has risen from about 9 per cent in 1965 to about 19 per cent in 1986.

This substitution in favour of clonal tea will continue so long as the farmers perceive – as they do – that clonal tea grows faster and produces better quality and yields more quantity of tea per land area. For the time being, however, the adoption rate and the adoption intensity of clonal tea must be reckoned to be highest in small farms and lowest in large farms. Most of our conclusions on comparative performance are based on this premise.

Tea output can be measured in monetary terms as the amount of money paid to the smallholder farmers by the KTDA which procures the tea output. It therefore excludes the processing and marketing costs which the KTDA deducts from its sales on behalf of its clients. Kenyan tea is sold in bulk at an auction market in Mombasa, Kenya, which is intimately linked to international commodity markets, particularly the London tea auction. To be comparable, tea output by the plantations, which have their own tea-processing plants and do their own bulk selling, was adjusted for factory and transport costs. Therefore tea output is, for our purpose, calculated as its value at farm gate.

Income generation and distribution

A glance at Table 5.6 indicates that output per hectare is largest in the case of small farms, suggesting the intensification of land use as farm size decreases. This is confirmed by the greater application of intermediate inputs – fertilisers, insecticides, etc. – as improved varieties of tea are grown. As expected, small farm operators use minimal capital *vis-à-vis* large farms, if only because workers in large estates tend to use sickles rather than nail-pick tea as is commonly done in small farms. Average wages per work-day range from Ksh.16 in large farms to Ksh.21 in small farms while profits and value added are greater in the small farm sector.

From the foregoing, it is clear that earnings by the factors of production, labour and entrepreneurship are superior in smaller farms, but that all cases compare very favourably with alternative viable undertakings, for example, coffee at Ksh.29 788 per hectare per year (Kenya (b), 1986). As in the case of potatoes, sectoral linkages (due to the stimulating effect of purchasing

TABLE 5.6 *Input, output, value added and profits (shillings per hectare)*

Indicator	Farm Size			
	Small	Medium	Large	All
Total output (1)	28 674	26 210	26 635	27 363
Intermediate inputs (2)	1 533	1 156	983	1 124
Equipment usage (3)	26	5	40	16
Labour costs (4)	3 665	6 380	9 363	5 604
Operatingcosts (2+3+4)=5	5 224	7 541	10 386	6 744
Profits (1–5)	23 450	18 669	16 249	20 619
Value added (1–2)	27 141	25 054	25 652	26 239

SOURCE: Survey data.

intermediate inputs) are strengthened and contribution to national income (value added) is increased by the introduction of improved clonal tea varieties.

The resultant income distribution is shown in Table 5.7 where the concentration indices – calculated as the share in income by the richest 30 per cent of the farmers relative to the share by the poorer 70 per cent – are given. With concentration ratios of between two and three for large and small farms, respectively, it would appear that the distribution of income is not disparate, though this conclusion is a bit weak when rapid adopters of improved tea varieties (small farm operators) are considered. To some extent, this consideration supports the Green Revolution effect observed earlier in the case of potatoes. It also supports the findings of a study covering the period 1974–75 that the Gini coefficient, at 0.59 for income from tea east of the Great Rift Valley in Kenya (mostly small-scale farms), is considerably higher than the Gini coefficient at 0.47 which obtains for tea incomes west of the Rift (where most of the

TABLE 5.7 *Distribution of income among tea farmers*

Percentage of income	Share of income in	
	Large estates	Small farms
Poorest (40)	11	10
Middle (30)	29	15
Richest (30)	60	75
Concentration ratio	2	3

SOURCE: Survey data.

large estates are located). These/calculations were done by J. L. Lijoodi and H. Ruthenberg (Clayton, 1983).

Impact on employment

From Table 5.8, we find that small-scale tea production is not particularly labour intensive – as measured either in terms of wages share in value added or in terms of work-days deployed per hectare of land. This could be because of the tendency for small farm operators to rely almost entirely on family labour, there being little labour available for hire as everyone in the neighbourhood is busy working on their shamba (farm, plot, parcel). Large estates, on the other hand, almost invariably have a standing hired labour force and access to casual workers who seek part-time or seasonal work. With the limited labour available to small farmers, they tend to spread it among the many farm activities such that the few allocated to tea work with deliberate speed so that the work-days expended are relatively few. Consequently, labour productivity is considerably higher in the small-scale tea farms than in the larger ones, and this is in spite of the almost negligible utilisation of capital equipment in tea farming. But looking at the proportion of value added to gross output, none of the farm sizes has an over-riding advantage from the point of view of this measure of relative efficiency.

Our analysis leads us to conclude that, while total factor productivity is rather high in the tea sector, and that in the small-scale farms land and labour are relatively more productive, labour intensity is considerably lower in small farms so that direct labour employment is unlikely to be augmented by the introduction of improved tea crop varieties to small farms.

From the preceding statement two corollaries come to mind. Firstly, there is a clear trade-off between employment creation and productivity.

TABLE 5.8   *Factor productivity and labour intensity*

| Indicator | Farm Size | | | |
	Small	Medium	Large	All
Wages share in value added (%)	13	25	36	21
Work-days per hectare	172	314	587	350
Capital use per work-day (Ksh.)	0.05	0.02	0.07	0.05
Labour productivity (Sh. per work-day)	167	83	45	78
Value added to gross output (%)	94	95	97	96

SOURCE: Survey data.

Secondly, the greater contribution to output by small-scale tea farms – per land area and per work-day – suggests that the indirect employment creation through the expenditure multiplier effect would be considerable, so that total employment creation would turn out to be far more with the introduction of improved varieties of tea than without.

CONCLUDING REMARKS

Improved varieties are unequivocally superior to traditional crops in terms of production and income generation. Our study revealed the mixed direct employment effect of modern varieties (MV), it being clearly higher in the case of potatoes but ambivalent in tea. However, when the indirect employment generated by increased income is taken into account, the total employment effect of MV crops exceeds the non-MV ones.

Comparing the two crops analysed, income per land area is greater for potatoes than for tea; so also is employment. But the resultant income distribution is more unequal in potato than in tea cultivation, though in both cases support is rendered to the contention that 'the introduction of MV technology into a community, in which resources are very inequitably distributed, tends to reinforce the existing inequality' (Hayami and Ruttan, p. 338). Tea provides a steady demand for labour all the year round while potatoes are grown in two seasons per year. Potatoes are more demanding of intermediate inputs but, being seasonal, can be operated on a 'get-in-get-out' basis – thus allowing for a quick entrepreneurial reassessment of the operational situation.

Almost universally, farmers complained of the high cost of certified potato seeds. While in the long term the Agricultural Development Corporation should find ways and means of lowering its sale price for seed potatoes, e.g. by more efficient operation and reduction of marketing margins, a credit scheme could immediately be put into practice for the acquisition of certified seeds, and other inputs. This could be facilitated by the formation of farmers' co-operative societies to deal in potatoes.

The marketing of tea is well handled by the Kenya Tea Development Authority for small farm operators and the Kenya Tea Growers' Association for the plantation sector. There is a need to streamline the marketing of potatoes, possibly through the co-operatives which could build storage facilities, foster canning of the crop and explore the possibilities of further processing into industrial and consumptive spirits, alcohol and starch.

Aside from serving as an animal feed (pigs, cattle), potatoes are an important source of carbohydrates, minerals and vitamins – thus serving as a vital component in national food security (Terry, *et al.*,1987). Tea is an important earner of foreign exchange. All these factors underline the socio-economic importance of both tea and potatoes.

The study furnishes concrete empirical evidence in support of the inverse relationship between farm size and productivity – both under the traditional and the new biotechnology in potato and tea. In all these cases, the small farms relative to large ones use less capital, make a larger contribution to national income and demonstrate stronger backward linkages to agricultural input suppliers.

Due to higher profitability and better financial capacity to bear the higher costs of investments associated with the new technology, large farms tend to pioneer the adoption of the new technology. It is also clear that the introduction of biotechnology into an agrarian system in which land is unequally distributed (the smallest 70 per cent of the Kenyan farms accounted for a meagre 10 per cent of the country's land area) tends to reinforce the existing inequality. In this respect the Kenyan experience is certainly a close parallel to the experience of inequality created by the Green Revolution in Asia.

This study, therefore, provides important evidence to the policy-makers about the developmental potential of small farm strategies both in a traditional agricultural setting and under dynamic conditions of technological change. In addition to growth, such a strategy enhances aggregate employment and promotes more equitable distribution of income. While this will help economise on the use of capital, it will help achieve higher output per unit of the increasingly scarce resource – land – given the current population explosion in Kenya. Clearly, the technocrats' implicit assumptions of the technological superiority of the large farms is unsupported by the evidence. Possibly this explains why the Kenyan Agricultural Development Corporation's potato project deals quite aggressively with the large-scale farmers, some of whom are used for the purposes of field multiplication of the certified (cloned) seeds.

6 Biotechnology and Labour Absorption in Malawi Agriculture

C. Chipeta and M. W. Mhango

INTRODUCTION

This chapter utilises *ex post* survey data to examine the effects of biotechnologies on production, incomes, employment, imports and exports in the tea industry in Malawi. For the purpose of the study three levels of technologies are considered: *Indian hybrid tea* denoted by T_1 represents tea plants with moderate sized leaves and with soft and large shoots; *polyclonal tea* denoted by T_2 represents an improved type of tea that is developed through cross-breeding and selection and propagated by growing tea bushes from seeds; and *clonal tea* denoted by T_3 also represents an improved type of tea that is developed through cross breeding and selection and propagated by growing tea bushes from leaf cuttings.

METHODOLOGY

The data analysed in this section was obtained from eight tea estates which completed and returned the questionnaire. As such, the material constitutes a case study rather than a random sample. Moreover, most of the eight estates were not able to fill in the questionnaire completely. Thus, the data in the tables may not reflect information from all the eight estates.

Two measures of tea output are used in this study. One is made tea measured in metric tonnes. This has been included to highlight actual physical yields of tea against expected yields per unit of land. The other is the gross value of tea output measured at constant prices. For clonal tea, output has been valued at the average market price realised between October 1986 and September 1987 as reported by the tea estates. The same is the case for Indian hybrid tea. For polyclonal tea, output has been valued at the average price realised between July 1986 and July 1987 as reported by the Smallholder Tea Authority. Because of wide disparities in

94

the prices of the various tea species, yields measured in monetary terms make more sense when considering the economic viability of improved species of tea.

Land (N) is defined in terms of area (measured in hectares) along with the tea plants growing within the confinement of that area. All the districts where tea is grown in Malawi meet the soil and climatic requirements for its cultivation. Whatever differences exist in the quality of the soil and in climatic conditions are believed to be insignificant and of little consequence.

The number of tea plants in a hectare varies from estate to estate. For all eight estates, the range of variation was from 6817 to 10 000 tea bushes per hectare. Within each tea species, the range of variation was smaller. For Indian hybrids, the range of variation was from 8430 to 9250; for polyclonal tea, it was from 6817 to 8197; and for clonal tea, it was from 8269 to 10 000 per hectare.

The density of tea plants per hectare depends on the spacing design. In the early years of the tea industry, the most popular spacing design was four feet by four feet square. Subsequently, other designs including hedge planting, gained popularity. Changes in the spacing patterns may have been induced by the expectation of better yields. However, it is doubtful that a larger number of tea plants per hectare will necessarily be associated with a higher yield per hectare.

Labour (L) is measured in work-days as reported by the estates. There is no disaggregation into male, female or adolescent labour. Total labour input is the aggregation of labour input in each operation. For tea estates which grow polyclonal and clonal tea, labour is inclusive of work days in the nursery and in new planting.

The intermediate material inputs (I) whose value has been totalled are fertilisers, chemicals for plant protection (pesticides, insecticides, fungicides and weedicides), fuel and power and others, including baskets, bags, etc. While fixed capital (K) represents depreciation of physical assets and the maintenance of the same.

Analysis of production

Technology levels
For the purpose of this study, technology level is associated with the type of tea that is grown. Indian hybrid tea will be denoted by T_1, polyclonal tea by T_2 and clonal tea by T_3. Not only is the average yield per hectare of the various species of tea different, the input combinations also differ. The distribution of estates by their technology level and their average

TABLE 6.1   *Number of tea estates by level of technology and by average size*

Level of technology	Number of estates using technology in hectares	Average size of land under each technology
T_1	7	284
T_2	5	53
T_3	6	56

size is shown in Table 6.1. The change in technology on tea estates almost entirely represents the variation in the application of biological innovations. As such, one must expect to find an increase in the use of intermediate material inputs per hectare (i.e. an increase in the coefficient I/N as intermediate material inputs are substituted for land). Ideally, this should be associated with an increase in land and plant productivity (O/N). As regards yields measured in monetary terms, this is true only in respect of clonal tea (see Table 6.2).

TABLE 6.2   *Intermediate material inputs, gross physical output (metric tonnes) and gross value of output per hectare by level of technology*

Level of technology	I/N (Malawi Kwacha)	O/N (metric tonnes)	O/N (Malawi Kwacha)
T_1	307.70	2.33	4 683.30
T_2	330.98	1.54	4 406.40
T_3	444.87	1.75	5 040.00

Average physical yields of clonal tea are below the 2 metric tonnes per hectare that the Tea Research Foundation expects from unirrigated land and far below the 5 metric tonnes per hectare it expects from irrigated land. The same is the case for polyclonal tea. Actual field yields are below potential for a number of reasons. Many polyclonal and clonal tea bushes are relatively young, hence they have not reached their maximum production potential. Clonal tea bushes are reported to be taking a long time to mature. Besides, many estates are not applying intermediate inputs, such as fertilisers, in sufficient amounts (this being a problem of managerial practices).

Smallholder tea is grown on three types of land: public land consisting of all land that was made available by government from virgin land, forest reserves or former estate land; customary land that was not used prior to the cultivation of tea, was fallowed for a long period of time or was virgin land granted by village headman; and customary land which was used for growing other smallholder crops. The mean yield is highest on public land and lowest on used customary land. At the beginning of the 1970s, these yields were estimated as 1049 kg per hectare on public land, 691 kg per hectare on unused customary land and 567 kg per hectare on used customary land. These yields refer to made tea. The higher yield on public land is attributable to the higher nutrient value of the soil and the good quality of the clearing work by the staff of the Smallholder Tea Authority.

Overall, annual smallholder yields of polyclonal tea are always below targets set by the Smallholder Tea Authority (see Table 6.3). These should be explained in terms of the untimely performance of certain farm operations and the low level of crop husbandry, particularly the low quality of weeding and plucking (Malawi Government, 1986). In 1984/85, for example, as much as 22.3 per cent of the green leaf plucked in Thyolo District was unacceptable. In Mulanje District, the unacceptable leaf was smaller at 6.9 per cent. The average for the two districts was 14.6 per cent. As between different years, variations in yields to a large extent reflect differences in the amounts of rainfall. Between 1975/76 and 1984/85 the initial fall in yields was due to inadequate price incentives. These had subsequently improved, hence the reversal in the yield trend.

Despite the higher average price of 264 tambala per kg applied to estate polyclonal tea as against 219 tambala per kg applied to Indian hybrid tea, the gross value of the former per hectare is smaller than that of the latter. Only the much higher average price of 319 tambala per kg for clonal tea makes a difference, despite the lower physical yield of clonal tea compared with Indian hybrid tea. Simultaneously, given the importance of the knowledge of applying the inputs of bio-logical innovations, one should expect to find more of these innovations

TABLE 6.3   *Smallholder green leaf yields compared with targets in kilograms*

Season	Target	Actual	Actual as % of target
1975/76	2 394 453	2 301 620	96.1
1976/77	3 165 566	2 887 508	91.2
1977/78	4 058 987	3 305 584	81.4
1978/79	5 032 590	4 040 117	80.3
1979/80	6 368 488	5 216 937	81.9
1980/81	7 421 161	5 923 101	79.3
1981/82	8 486 109	6 079 338	71.6
1982/83	9 526 614	6 985 937	73.3
1983/84	10 513 687	7 757 201	73.9
1984/85	11 298 100	10 208 975	89.6

SOURCE: The Smallholder Tea Authority Annual Report 1984–85.

TABLE 6.4   *Average factor input (land, labour, intermediate material inputs and capital flow services) per metric tonne of tea output by level of technology*

Level of technology	Land in hectares	Labour in man-years	Intermediate material inputs (K)	Capital flow services (K)
T_1	0.37	0.85	256.47	97.86
T_2	0.66	1.77	220.16	130.67
T_3	0.60	1.59	259.17	118.27

to be associated with an increase in output per hectare. The principal manifestation of this association would be seen in the economy in the input-use per unit of output (i.e. biological innovation substituting all or some traded inputs). This economy could take two forms: (a) a reduction in the total factor cost per unit of output where less of one factor but more of another is used, and (b) a reduction in the total factor cost when less of each factor per unit of output is used. If the comparison was between T_2 and T_1 levels of technologies, for example, T_2 would be more profitable than T_1 in the first case but superior in the second.

In Table 6.4, quantities of land, labour, intermediate material inputs and capital flow services per unit of output are presented. It can be seen that the inputs of land rise as one moves from the T_1 to the T_3 level of technology, reaching its peak in the T_2 level of technology, and similarly for labour and capital inputs. The only partial exception to this general pattern occurs in the case of intermediate material inputs. Here there is a decline between T_1 and T_2, but the input at the T_3 level of technology exceeds the inputs at each of the T_1 and the T_2 levels of technology. What these facts tell us is that there is a shift in the production function as one moves from T_1 to T_2 and T_2 to T_3 levels of technology. Since output is measured in physical units, we would tend to conclude that the path of change in technology is from superior to inferior techniques. If output were valued in monetary terms, we would probably conclude that the change in technology was from inferior to superior techniques as far as clonal tea is concerned.

Analysis of variance

In order to assess the relative importance of different factors to changes in technology, an analysis of variance is attempted. Estates are classified by T_1, T_2 and T_3 levels of technology and the degree of variation of the variable listed below is tested.

A.	(i)	Work-days of labour per hectare	(L/N)
	(ii)	Intermediate material inputs per hectare	(I/N)
	(iii)	Capital flow services per hectare	(K/N)
B.	(i)	Size of estate	
	(ii)	Number of tea bushes per hectare	(B/N)

By taking each input per unit area of land in A, we abstract from the scale of operation. Land area being homogeneous, we can see the variations in each input in combination with the same amount of land. In B, we test whether the size of estate varies significantly with change in technology. We also test whether the number of tea bushes per hectare varies with change in the level of technology.

Results of the analysis are presented in Table 6.5. It can be seen that L/N, I/N and size of estate vary significantly with the level of technology. We can, therefore, conclude that in the type of technology by T_1, T_2 and T_3, capital flow services per hectare (K/N), and the number of tea bushes per hectare (B/N) are not significantly different.

Regression Analysis

Through the following regression analysis, the significance of the species of tea grown is tested. A log linear regression model of the form:

$$\log y = a + b_1 \log x_1 + b_2 \log x_2 + \cdots \cdots + b_n \ \log x_n$$

in which the proxy variable P represents technology level or species of tea grown will be used. P will take the value 1, 2 and 3 for estates in the T_1, T_2 and T_3 levels of technology.

Four functions are presented:

  (i)   O/N = f(I/N, L/N);
 (ii) O/N = f(I/N, L/N, L/N, B/N);
(iii) O/N = f(I/N, L/N, P); and
(iv) O/N = f(I/N, L/N, K/N, P).

These functions show (Table 6.6) that intermediate material inputs and the number of bushes per hectare are highly significant in explaining variations in output. Labour and the level of technology are also significant in some of the equations.

TABLE 6.5   Results of analysis of variance for testing the effects of technology change on a number of selected variables

Variable	Sum of squares due to technology	Degrees of freedom	Mean sum of squares due to technology	Sum of squares due to error	Degrees of freedom due to error	Mean sum of squares	Variance ratio F
Labour per hectare	95552.82	2	47776.41	33663.26	15	2244.22	21.29***
Material input per hectare	65718.54	2	32859.27	85049.09	15	5669.94	5.80***
Bushes per hectare	12200688.00	2	6100344.00	0.00	15	0.00	0.00
Size of estate	105003.34	2	52501.67	7.17	15	0.48	109785.09***
Capital flow services per hectare	0.05	2	0.03	77865.63	15	5191.04	0.00

*** Significant at the 1 per cent level.

TABLE 6.6 *Regression outcomes of alternative functional forms*

Explanatory variables[1]	Regression coefficients[2]			
	Function I	Function II	Function III	Function IV
Intermediate material inputs per hectare (I/N)	0.37046*** (0.12244)	0.29719*** (0.11381)	0.38181*** (0.12812)	1.20455*** (0.16365)
Labour (work-days) per hectare (L/N)	-0.21678 (0.34432)	0.45595 (0.41192)	-0.03225 (0.57709)	1.26010*** (0.40547)
Bushes per hectare (B/N)	–	1.02947** (0.41703)	–	–
Capital flow services per hectare (K/N)	– (0.13133)	–	–	-0.47884
P as proxy for technology	–	– (0.08076)	-0.03255 (0.04951)	-0.12325**
Constant	7.74517*** (2.19923)	-5.55474 (5.73757)	6.51314* (3.79193)	-4.30294* (2.77630)
R^2	0.23817	0.38689	0.20653	0.79588

*   Significant at 10 per cent level.
**   Significant at 5 per cent level.
*** Significant at 1 per cent level.
1.   I = intermediate material inputs; N = area cultivated in hectares;
       L = labour in man-days; B = number of tea bushes; K = capital cost.
2.   Figures within brackets are standard errors of the estimates.

TABLE 6.7   *Change in factor coefficients I/N, L/N and I/L*
at different levels of technology

Factor coefficients	T_1	T_2	T_3	Change at T_2 level compared with T_1 level	Change at T_3 level compared with T_2 level	Change at T_3 level compared with T_1 level
I/Na	514.43	330.98	444.87	−183.45	+113.89	−69.56
L/N (Work-days)	671.51	828.70	837.11	+157.19	+8.41	+165.60
I/La	0.78	0.40	0.53	−0.38	+0.13	−0.25

SOURCE: In Kwachas (local currency).

Factor proportions

Changes in factor proportions or movement along the production function are shown in Table 6.7, in terms of coefficients I/N, L/N and I/L. Labour inputs do not apply to fertilising because this operation is carried out by aerial spraying.

Intermediate material inputs per unit of land are smaller at each successive level of technology than at the T_1 level. There is, however, an increase between the T_2 and T_3 levels. This is also the case for intermediate material inputs per unit of labour. The lower levels of intermediate material inputs per unit of land or labour at the T_2 and T_3 levels of technology as compared with the T_1 level is attributable to a number of factors.

First, the level of fertiliser application at the T_2 level is much less than at the T_1 level. This might mean that polyclonal tea is less receptive to larger amounts of fertilisers or that the right amount of fertiliser is simply not applied. At the T_3 level of technology, the amount of fertiliser used per hectare is not significantly different from the one used at the T_1 level, but it is certainly larger than the amount used at the T_2 level.

Secondly, the amount of other chemicals used up per hectare is lower at both the T_2 and the T_3 levels than at the T_1 level. The chemicals concerned are those used in plant protection and include pesticides, fungicides and herbicides. Polyclonal and clonal tea plants are more resistant to diseases than Indian hybrids. Therefore, they require fewer material inputs for their protection.

Thirdly, the amount of fuel/power required per hectare of polyclonal and clonal tea also is lower. This may be explained in terms of the smaller amount of plant protection required at the T_2 and T_3 levels of technology and the generally smaller number of tea bushes per hectare of tea at the same levels of technology compared to the T_1 level of technology.

Lastly, the amount of other material inputs (including baskets, bags and other packaging materials) required per unit of land at the T_2 and T_3 levels of technology is around half of the amount required at the T_1 level. This has to do with the lower physical yields of polyclonal and clonal tea per hectare.

At all levels of technology proportionately more of the expenditure on intermediate material inputs goes into fertilisers, fuel and power than into other chemicals and other intermediate material inputs. Compared to T_1 and T_3, the proportion spent on fertilisers at T_2 is less, while the proportions spent on fuel and power and other chemicals are larger (see Table 6.8).

The labour input per unit of land increases as one moves from the T_1 to T_2 and from T_2 to T_3 levels of technology. Whereas virtually no labour is required for Indian hybrid nursery and planting work, some is required in connection with the establishment of polyclonal and clonal tea. Besides, the pruning of polyclonal and clonal tea bushes which are relatively young requires more labour per hectare than the pruning of Indian hybrid tea bushes. Otherwise, less labour is required per hectare in weeding, protecting and in plucking polyclonal and clonal tea.

In all comparable operations, labour cost per hectare of polyclonal and clonal tea is much less than per hectare of Indian hybrid tea. In weeding, for example, it is K3.48 as against K15.00. And in plant protection, it is K5.00 and K8.00 in polyclonal and clonal tea, respectively, as against K9.00 in Indian hybrid tea. The planting of Indian hybrid tea is more widely spaced than that of other species and accounts for the observed difference in labour costs.

TABLE 6.8 *Proportion of expenditure on components*
of intermediate material inputs per unit of land in tea cultivation
by levels of technology (per cent)

Intermediate	Levels of technology		
	T_1	T_2	T_3
Fertilisers	69.9	64.1	72.4
Other chemicals	7.9	10.6	7.8
Fuel and power	18.3	22.6	17.7
Other intermediate material inputs	3.9	2.7	2.1
Total	100.0	100.0	100.0

Analysis of Incomes

Direct income effects
The average yield of Indian hybrid tea on the case study estates for the six year period 1980/81 to 1985/86 was 2.33 metric tonnes of made tea per hectare. Corresponding yields for polyconal tea and clonal tea were 1.54 metric tonnes and 1.75 metric tonnes respectively. On these estates in 1985/86, this implies that total production was 211.19 metric tonnes lower for polyclonal tea and 193.44 metric tonnes lower for clonal tea than what it would have been if all the land had been devoted to Indian hybrid tea. Altogether, tea output was 403.63 metric tonnes lower.

According to our estimates, the loss in revenue arising from this was K270 989.44 in respect of polyclonal tea. This was only partially offset by a gain of K160 046.40 for clonal tea. Hence, a net loss of K110 943.04 was sustained in revenue. In terms of value added, the net loss was K35 426.81.

As far as wage income is concerned, there is no difference in wage rates between workers depending on the species of tea on which they work. (There may in fact be little differentiation of work by species of tea.) The average wage rate applicable to manual workers on all tea estates in 1985/86 was 95.2 tambala per work-day. The 1 189 498.2 work-days attributable to Indian/local tea on six of the case study estates earned around K1 132 402.29 in cash in 1985/86; while the 175 312.8 work-days attributable to polyclonal tea earned around K166 897.79 and the 99 060.0 work-days attributable to clonal tea earned around K94 305.12. Since polyclonal tea requires 21.9 per cent more labour per hectare than Indian/local tea, if all the land was under polyclonal tea, wage income of K1 132 402.29 would have been up by K247 996.10. On the other hand, if the land under Indian/local tea had all been under clonal tea, the relevant wage income would have been up by K260 452.53 since clonal tea requires 23.0 per cent more labour per hectare than Indian/local tea. Information on marginal propensities to consume is not available to enable us to estimate the indirect effects of these changes on total income through the multiplier process.

Since polyclonal and clonal tea are grown on land which should have been under Indian and local tea, the net gain in employment is only 41 023.2 work-days for polyclonal tea and 24 467.6 work-days for clonal tea in respect of the case study estates. The corresponding net gains in wage income are K39 054.08 and K23 293.36 for polyclonal tea and clonal tea respectively.

Research related to the development of new tea biotechnology is a significant source of income in the country. For example, the 933 people that were employed by the Tea Research Foundation in 1986 earned some K704 371 between them.

Indirect income effects

As a result of the reduction in imported material input requirements, value added in the distribution sector was lower by K96 172 in 1985/86. Of this amount, K27 382 represented wages and K63 381 represented net profit. Malawi does not manufacture fertilisers and chemicals which are the most important imported material inputs. Hence it is not necessary to estimate the net impact of their domestic manufacture.

Analysis of employment

Direct employment effects
In 1985/86, manpower requirements per hectare in field operations were 70.8 work-days in growing Indian/local tea, 86.34 work-days in growing polyclonal tea and 87.11 work-days in growing clonal tea. In other words, polyclonal and clonal tea field manpower requirements were greater than those of Indian/local tea by 15.54 work-days or 21.9 per cent and 16.31 work-days or 23.0 per cent respectively. If we are to apply to these figures the total hectarage under each type of tea in 1985/86, we obtain employment gains due to polyclonal tea of 45 394 work-days and due to clonal tea, 26 846 work-days.

Inputs which displace labour are used in operations carried out for plant protection. For example, chemical weedicides for controlling weed growth and power sprayers for spraying pesticides, insecticides or fungicides substitute labour directly. As there are various types of weeds and methods of eradicating them, the amount of labour used in weed control at any level of technology depends on the intensity of the weed control activity and on whether there is more use of manual weeding than chemical eradication. Which method is adopted depends on the intensity of weed growth and the type of weeds that require the more intensive approach. It also depends on the approach to plant protection that is followed, that is, on the importance given to plant protection and especially to weed control in the organisation of operations and in the scheme of operationwise resource allocation.

The data shown in Table 6.8 indicate that there is a net increase in labour input per hectare in nursery and planting between the T_2 and T_3 levels of technology. In other words, the development of clones and their establishment in the field requires more labour than the development and establishment of polyclonal tea. At present, there is no planting of Indian hybrid tea. Hence the large increase in the labour input per hectare in nursery and new planting between the T_1 and T_2 and the T_1 and T_3 levels of technology: in pruning, there is a net increase in the amount of labour input per hectare between T_1 and T_2 and the T_1 and T_3 levels of technology because of the relatively younger age of clonal and polyclonal tea bushes, which consequently require more frequent pruning than Indian hybrid tea bushes. However, between the T_2 and T_3 levels of technology, there is a net decrease in the amount of labour inputs per hectare in pruning implying that clonal tea bushes require less frequent pruning than polyclonal tea bushes.

In weeding, there is again a net decrease in the amount of labour input per hectare between the T_1 and T_2 and the T_1 and T_3 levels of technology implying that there is more use of chemicals in controlling weeds at the T_2 and T_3 levels than at the T_1 level. Between T_2 and T_3 there is, however, a net increase in labour input per hectare in weeding. In plant protection, the pattern is similar, a net decrease in labour input per hectare between the T_1 and T_2 and the T_1 and T_3 levels of technology because of greater use of mechanical means at the T_2 and T_3 levels of technology, and no net change in labour input per hectare between the T_2 and T_3 levels of technology. So too in plucking, a net decrease in the labour input per hectare between the T_1 and T_2 and the T_1 and T_3 levels of technology because of lower yields of tea at the T_2 and T_3 levels of technology, and a net increase in labour input per hectare between the T_2 and T_3 levels of technology because of a higher yield of tea at the T_3 level of technology compared to the T_2 level. Considering all operations together, there is a net increase in the labour input per hectare between the T_1 and T_2 and between the T_2 and T_3 levels of technology, as well as a net increase in the labour input per hectare between the T_2 and T_3 levels of technology.

Organisational improvements

Organisational improvements in the use of labour can come about through general rationalisation of labour and improvement in supervision in each operation through reallocation of labour between operations in such a way that the total input of labour per unit of output is reduced. Both ways may be used simultaneously.

With respect to the latter, one can see from Tables 6.9 and 6.10 that for the case study tea estates as a whole, there is a proportionate as well as an absolute decline in the labour input per hectare in weeding, plant protection and plucking between the T_1 and T_2 and also between

TABLE 6.9   *Change in labour (work-days) input per hectare in different operations by level of technology*

Operations	Between T_1 and T_2	Between T_2 and T_3	Between T_1 and T_3
1. Nursery and new planting	+18.3	+6.8	+25.1
2. Pruning	+15.2	−15.4	+0.1
3. Weeding	+6.8	+2.8	−4.5
4. Plant protection	−1.9	0.0	−1.9
5. Plucking	−9.6	+7.1	−2.5
Total	+15.2	+0.8	+16.3

the T_1 and T_3 levels of technology. In nursery and new planting there is a proportionate as well as an absolute increase in labour input between T_1 and T_2, T_1 and T_3 and T_2 and T_3 levels of technology. In pruning, the proportionate and absolute increases in labour input per hectare is between the T_1 and T_2 levels of technology only; between T_1 and T_3 (only in proportional terms) and between T_2 and T_3 there is a decline. These differences are attributable to variations in yields and in cultural practices at different levels of technology.

Pruning, in terms of labour input, is clearly the most important activity in tea, accounting as it does for over half of the total labour input. The direction in which labour input per hectare for pruning changes with an advance in technology very largely determines the change in the total labour input in tea estates. Estates do not seem to have made a systematic effort to rationalise the labour input in this activity. The data in Table 6.9 indicate that an advance in technology between T_1 and T_2 levels is accompanied by more labour input per hectare in pruning, but an advance in technology between T_2 and T_3 levels of technology is accompanied by a reduction in the labour input per hectare in the same activity.

TABLE 6.10   *Percentage distribution of labour input (work-days) per hectare by operations and level of technology*

Operations	Level of technology		
	T_1	T_2	T_3
1. Nursery/new planting	0.0	21.2	28.3
2. Pruning	51.7	60.3	42.1
3. Weeding	17.4	6.4	9.0
4. Plant protection	4.2	1.3	1.3
5. Plucking	26.7	10.8	18.8
Total	100.0	100.0	100.0

Indirect employment effects of technological change

As we have pointed out already, direct employment measured in work-days increases from 671.51 at the T_1 level of technology to 828.70 at the T_2 level of technology and 837.11 at the T_3 level of technology. This includes labour in all field activities connected with tea cultivation, including nursery work and new planting. The establishment of clonal tea is a long-term process that will last into the next century. As such, it has become a permanent feature of tea cultivation. This justifies its inclusion in manpower data.

Indirect employment attributable to a change in technology arises in three main ways: through manufacturing activities where tea is used as an input, through a forward linkage induced by an increase in income and consisting of additional saving and investment and additional consumption and expenditure; and through a forward linkage with industries which provide the additional non-labour inputs required for the new technology.

Since the net effect of introducing polyclonal and clonal tea is to reduce the output of made tea, the change in biotechnology has the effect of reducing employment in tea manufacturing. It has been estimated that for the case study estates, the loss in employment in tea manufacturing arising from a net reduction in made tea of 440.63 metric tonnes was 36 jobs.

Employment generated by savings and consumption induced by additional wage income

An advance in technology from the T_1 level to the T_2 and T_3 levels will cause income to increase because it is accompanied by an increase in the amount of labour required per hectare. The wage rate applicable at the various levels of technology is the same and, hence, it does not account for differences in incomes at different levels of technology. Given a rise in income, it is reasonable to believe that there would be an increase in savings and investment. Unfortunately, it has not been possible to estimate the increase either in savings or in investment. Therefore, estimates of the indirect employment that might result from this source, which may not be negligible, are not presented.

Additional consumption expenditure due to a change in technology constitutes additional demand for consumer goods. When this is met through increased productivity, this causes additional labour to be employed and additional income to be generated. The second round of consumption expenditure starts another round and so it goes on until the original impetus is exhausted. The initial increase in consumption expenditure in this case arises out of the difference in all factor incomes at the T_2

level of technology compared with T_1 and at the T_3 level of technology compared with T_1. We shall limit ourselves in this study to increases in wage income.

When the income of an individual or community rises over time, the additional consumption depends on the marginal propensity to consume (MPC). But when comparison is being made between two individuals or groups, having different levels of income at the same time, additional consumption attributable to the difference in their incomes will be given by their respective average propensity to consume (APC).

Because of a lack of data on marginal and average propensities to consume in respect of tea estate workers, on the division of expenditure between products of agriculture or rural industries and on the products of urban industries, it has not been possible to estimate the indirect income and employment resulting from the additional income and consumption expenditure at the T_2 and T_3 levels of technology compared with the T_1 level.

In addition, as a result of a reduction in imported material input requirements, it is estimated that employment in the distribution sector was 37 work-years below what it would be if only Indian hybrid tea, which requires more imported .naterial inputs than polyclonal tea, was grown.

Analysis of imports and exports

On the basis of the information contained in completed questionnaires, the value of imported inputs in 1985/86 has been estimated at K564.95 per hectare of Indian hybrid tea, K396.98 per hectare of polyclonal tea and K546.10 per hectare of clonal tea. This implies a saving of K167.96 per hectare of polyclonal tea which is significantly larger than the saving of K18.84 realised per hectare of clonal tea. Applying the total hectarage under polyclonal and clonal tea in 1985/86 to these figures, gives total savings of K490 611.16 and K31 010.64 respectively. The main imported inputs to which these savings relate are fertilisers, other chemicals used in plant protection, fuel, power and others, mainly packaging materials. The amount spent on all these items per hectare of polyclonal tea is the lowest, followed by clonal tea except with regard to fertilisers where there is no significant difference between the amount used per hectare of clonal tea and Indian hybrid tea.

The reduction in imported material input requirements has the following important consequence. Demand for limited foreign exchange resources to purchase those inputs is curtailed. This is welcome in view of the limited resources of foreign exchange available to the country.

The average price realised at the London auction market of polyclonal tea grown by smallholder farmers is higher than the average price realised at the same market by all tea from Malawi. Between July 1984 and June 1985, the difference between the two was, on average, 24 pence per kg. The world average price during this period was only slightly higher than the average of Malawi polyclonal tea grown by smallholder farmers. The difference was only 4 pence per kg. Assuming that if the smallholder farmers whose tea was processed by the Malawi Tea Factory Company had not produced the 1 345 711 kg of made polyclonal tea, but 1 345 711 kg of Indian hybrid tea, foreign exchange earnings from the export of the crop would have been lower by 11.9 per cent.

There is no data on the average prices of the different types of tea that are grown by estates that would facilitate estimates of gains in export earnings arising from the change in technology from Indian hybrid tea to polyclonal tea or clonal tea. Only a handful of estates provided the relevant data. But given the diversity of markets (London, Limbe and others) and modes of marketing (auctioning, private treaty, etc.), it was felt that the data provided by a small number of estates would not be representative of the whole industry.

For the case study estates, if all land had been under polyclonal rather than Indian hybrid tea, export earnings would have been greater by 20.7 per cent in 1986/87. On the other hand, if all land had been under clonal tea rather than Indian hybrid tea, export earnings would have been larger by 45.7 per cent in the same year.

CONCLUDING OBSERVATIONS

The observed biotechnology used in Malawian tea is of the simple variety involving cross-breeding and selecting plan species with desired characteristics, such as high yield, improved quality and resistance to drought or diseases. Vegetative propagation rather than cloning by tissue culture is used to multiply the chosen varieties of clonal tea both at the research station and in the field.

Historical factors are more important than anything else in determining whether a particular tea planter is growing improved species or not. The scale of operation is not a crucial factor. Availability of land and financial resources constrain the further adoption of improved tea species. Estimates of future hectarage, production and yields (see Table 6.11) indicate that total hectarage will increase by 2.1 per cent only between 1986 and 1990. This reflects the land constraint to which we have already referred. Given

TABLE 6.11   *Estimated future tea production*

Indicator	Season ending in				
	1986 (actual)	1987	1988	1989	1990
Total tea area (hectares)	18149	18203	18241	18382	18533
Weight:					
Members' estimates (kg 000s)	33061	40684	41515	42818	43941
TAML[1] estimates (kg 000s)	42000	41000	42850	44100	45000
Average yield based on members' estimates: kg per hectare	2208	2235	2276	2309	2371
Indexed to 1964 = 100	191	194	197	202	205
Based on TAML estimates: kg per hectare	2147	2417	2535	2638	2572
Indexed to 1964 = 100	186	209	220	228	238

1. Stands for Tea Association of Malawi.
SOURCE: Price Waterhouse, 1985.

the lack of adequate unused land, lack of capital, low tea prices and high input and transport costs, the tea industry is unlikely to switch completely from local and Indian hybrid tea in the near future. Most tea estates seem to be replacing old with new varieties at a rate of 2 per cent per year. At this rate of replacement, it will take 45 years to complete the exercise.

Most of the land planted with tea by smallholder farmers has polyclones. These species were adopted by the Smallholder Tea Authority because of ease of establishment and transport. To replace all the existing hectarage with clones would be very costly in terms of the amount of new investment required and forgone production, incomes and foreign exchange earnings. Besides, farmers would have to learn how to handle new species, especially when they are in the nursery. Clonal tea is relatively new and one does not know how it is going to stand in the field, especially under smallholder conditions. The Smallholder Tea Authority has allowed farmers to fill in with clones and all new expansion is with clonal tea. However, for a variety of reasons, including lack of suitable land and other resources, expansion is limited. So it is unlikely that there will be a large switch to clonal tea in the near future.

The observed physical yield of estate polyclonal and clonal tea is lower than the physical yield of Indian hybrid and local tea. It is also lower than the yields of the same tea species expected by the Tea Research Foundation under field conditions. This phenomenon reflects the relatively younger age of the polyclonal and clonal tea bushes, the slow growth of the clonal tea species that have been adopted, application of inadequate material inputs and lack of irrigation. In order to raise yields and hence close the gap, tea estates will have to plant faster growing clonal tea species, apply more material inputs and use irrigation wherever possible. The physical yields of smallholder polyclonal tea are also lower than expected yields because of inadequate application of fertilisers and the low quality of soil and crop husbandry and plucking. The Smallholder Tea Authority must successfully grapple with these problems if yields are to rise.

In monetary terms, the yield of clonal tea is larger than the yield of Indian hybrid and local tea because of the higher market price of clonal tea. For this reason, it is in the best interests of tea estates to grow more clonal tea than Indian hybrid and local tea. In contrast, the yield in monetary value terms of polyclonal tea is smaller than that of Indian hybrid and local tea because of the lower market price of the former. For this reason, estates do not find it attractive to plant more polyclonal tea.

Therefore, a major policy concern arises out of the constraints faced by estates and smallholder tea growers in adopting clonal tea. These growers can cultivate clonal tea either by planting new land or by replacing other varieties of tea with clonal bushes. Given that the availability of suitable land is a big constraint in Mulanje and Thyolo Districts, further expansion should be encouraged in Nkhata Bay District where suitable land and irrigation water exist. Acceleration of the pace of replanting existing fields with clones is dependent upon the improvement in the financial environment. According to the International Monetary Fund, the prices of tea, which are currently depressed, should pick up. This will give tea growers some incentive but, in itself, will not be enough. Every assistance should therefore be given to the Tea Research Foundation to work on techniques of tea manufacture with a view to devising more efficient methods, while the tea industry must continue to improve management practices to raise yields and lower production costs. There is also a need to moderate the growth of imported input costs by refraining from devaluation of the currency. Furthermore, there is a need to make credit available on easy terms to those tea firms which would like to adopt clonal tea; and to review the disincentive effects of the income tax on foreign-owned firms and the tax on dividends in the hands of the recipient company.

The research and development (R&D) associated with improving tea species is a comparatively significant source of both income and employment in Malawi. According to the Tea Research Foundation, employment in the tea industry should rise with an advance in technology. The observations that we have made support this view with respect to the amount of labour required per hectare in nursery work, planting and pruning, but not in plucking, plant protection and in weeding. However, the net employment effect is positive. The indirect employment created by advances in tea biotechnology is negative both in tea manufacturing activities and in the distribution sector which supplies intermediate material inputs to the tea industry.

Differences in workers' earnings on tea estates by level of technology have not yet been observed. This implies that the differences in workers' earnings on estates growing different tea species are due to differences in the numbers of workers and not the wage rates.

As polyclonal and clonal tea require fewer imported inputs and fetch higher prices on the world tea market, their contribution to the country's balance of payments is positive. However, the gain from higher export prices of polyclonal and clonal tea is not being fully realised. Several tea estates, with small amounts of clonal tea, mix it with the larger quantities of Indian, local and polyclonal tea during manufacture. As a result, they do not benefit from the prices which higher quality clonal tea would fetch on the world market. The net balance-of-payments effect could therefore be negative. With polyclonal tea grown on estates, the net balance of payments effect could also be negative. As far as smallholder polyclonal tea is concerned, the story is different. This type of tea requires fewer imported inputs per hectare than clonal tea. Because the Malawi Tea Factory Company is able to manufacture pure polyclonal tea which fetches higher prices than Indian hybrid or local tea, the impact of smallholder polyclonal tea on the balance of payments must certainly be positive.

Part III
Milk and Feed

7 The Differential Impact of Biotechnology: the Mexico – United States Contrast

Gerardo Otero

INTRODUCTION

The purpose of this chapter is to examine the impact of two specific new biotechnologies, the high fructose corn syrup (HFCS) and the bovine growth hormone (BGH), on two totally different socioeconomic settings – Mexico and the United States. In the case of the latter, it is nearly an *ex post* assessment of the socioeconomic impact, while in the case of the former it is more of an *ex ante* analysis of possible impact as these biotechnologies have not reached the application stage.

The bovine growth hormone (BGH) relates to milk production, and high fructose corn syrup (HFCS) is a substitute for sugar from cane and sugar-beet. Both of these products are already in the American market. HFCS entered the market in the late 1970s, and BGH was approved for marketing in 1988.

It can be said that the differential impact of these technologies in general will be as follows. While HFCS will substitute much of the formerly imported sugar to the United States from Third World countries, it may not directly affect employment in that country. Conversely, the introduction of BGH will probably eliminate about 50 per cent of dairy farmers in the United States and many workers in the industry, given the usual milk overproduction in the country over the past several years. On the contrary, BGH may have a positive impact in Mexico, where the demand for milk currently exceeds national supply. On the other hand, should the use of HFCS be expanded in Mexico, the effect on the hundreds of thousands of people who depend on the sugar industry will be devastating, while the consumers might benefit. The context and issues involved in the introduction of BGH in the United States and Mexico are first taken up, followed by discussion on HFCS.

117

BOVINE GROWTH HORMONE

Polarisations in the United States

Cornell researchers demonstrated milk production increases up to 40 per cent with BGH-treated cows. Costs also increase, along with food intake, but they do so at a lower rate: 'Overall, with stable milk prices, farm returns over variable costs increase by 5 to 26 per cent depending on farm characteristics and the response of animals to hormone administration' (Kalter, 1985).

The secret of BGH seems to be feed efficiency. Citing Dale E. Bauman, the director of the pioneering research, Melanie DuPuis says that the hormone does not affect the cow's maintenance requirement. And because about 30 per cent of a cow's feed goes into maintaining her own body, 'producing extra milk without extra maintenance feed greatly increases efficiency'. Increased yields vary depending on the lactation stage of the cow. Treating cows 60 days after calving increased milk yields in Bauman's tests by 10–15 per cent. But in the late stage of lactation yields averaged 30–40 per cent above normal. 'The average increase in milk over the entire lactation cycle was 25 per cent' (Dupuis, 1985 and Sun, 1986).

According to studies by Cornell agricultural economists and others, it is believed that a 100 per cent adoption rate for BGH could take place in a period of three years. Some of the consequences of such a fast adoption rate will be that: non-adopters will survive, surplus lands will be released, and a 30 per cent reduction in the United States herd of cows will result by the year the 2000. These changes will also involve greater pressures on United States government policy for supply control.

Given past experience of technology adoption by the American farmers, it could be expected that they will adopt BGH in a three year period. Also, as milk prices and the herd size drop with the use of BGH, less land will be needed for dairying. Thus, the poorest lands will become redundant, and they could possibly be put to good use with the aid of other agricultural biotechnology products which become available (Kalter, 1985). A financial side effect of this is a further decline in land value in the United States, which will undercut the ability of farmers to use land as collateral for credit (Kalter and Magrath, 1985). Land values have dropped severely already during the 1980s, placing a tremendous burden on farmers, many of whom have gone bankrupt.

Moreover, Lewellyn Mix (1986) has predicted a 30 per cent cow reduction by the year 2000. Based on a number of plausible assumptions regarding population growth, growth in milk consumption and the rate of technology adoption, Mix's paper contrasts several estimates of future production based on whether BGH is adopted or not. If it is not, demand would still be satisfied by the year 2000, assuming a yearly yield increase based on other genetic and managerial improvements. Such yield increase will account for 47.3 per cent of total increase if BGH is adopted fully by 1998, and the latter will account for the remaining 52.7 per cent of total yield increase (Mix, 1986).

Because of the mixed effects expected from BGH, controversy arose in the United States over its introduction. Traditional opponents of biotechnology, like Jeremy Rifkin, allied with farmers' organisations and the Humane Society, which defends animal rights.

Nevertheless, controversy and debate have not been enough. It depends on who ultimately controls the decisions on research policy and funding. If such control is increasingly in industry's hands, no matter what the public thinks, industry will always put efficiency and profitability ahead of public interest. And, in this case, BGH has already been approved for marketing in the United States. It is also interesting that this controversy over BGH introduction had the effect of preventing adoption to a large extent. This is one important example of how public concerns can have a decisive influence on the adoption of a new technology.

It is believed that the United States could have 30 per cent or 3.3 million fewer cows, 51 per cent or 92 500 fewer dairy farms, 195 000 fewer dairy farm employees, and use 9–10 million fewer crop acres for dairy feed production. The expected *rate of change*, not the direction, is of greatest importance and concern. Dislocation is believed to be inevitable (Mix, 1986).

The above scenario is shared by most analysts in the United States. Although farmers who survive with the new BGH will better compete with foreign dairy farmers and domestic producers of alternative beverages and foods, on the whole, new biotechnologies, including BGH, will deepen the now-longstanding pattern of structural change in American agriculture towards fewer and larger farms and lower commodity prices. The reasons are that new technologies require more expensive and capital-intensive inputs, and they are management intensive (Buttel, 1986).

According to Buttel, the most vulnerable group to shifts in technological forces are medium-sized family farmers, and even some larger farms with very high debt-asset ratios. Small farms continue to be viable, and even experience some growth since the early 1970s, thanks to the subsidy of non-farm incomes. Conversely, the greatest beneficiaries are agribusiness input firms, consumers and early adopters (mostly large farmers).

Along with these changes in agrarian structure, changes in agricultural research have taken place in the United States. Indeed, the latter changes have led to increased use of purchased inputs. The shift was from a predominantly publicly-funded to a predominantly privately-funded agricultural research. Currently, public agricultural research institutions account for only one-third of R & D expenditures, and private firms account for two-thirds. This has meant that, over the past two decades, new technologies have been proprietary, thereby assuming the character of purchased inputs. The introduction of BGH will reinforce such trends.

Economies of scale are clearly biased towards larger farms. This was determined by Kalter's study of three representative farms: 'On a per cow basis, increased return is the greatest on the large farm . . . and decreases progressively with farm size. Likewise, the increase in return per hundredweight of increased milk production increases with farm size'. Moreover, management skills are crucial: 'farm management ability of individual operators will be absolutely critical to the successful economic use of a product like BGH' (Kalter, 1985: 6).

Increased skill requirements will also lead to greater need for use of other technologies such as microcomputers. Kalter's predictions on management skill requirements for good use of BGH has been supported by other analysts. It is stressed that to maximize production, dairy farmers will rely increasingly on computers to keep exact records of a cow's milk output and to monitor its feeding requirements (Sun, 1986).

All of these arguments confirm the view that BGH will enhance the need for a 'systems approach' to farming, with increased requirements for ever more expensive purchased inputs, more or less conforming 'packages'.

According to a study of BGH impacts on New York State, consumers will be the primary beneficiaries. Assuming a free market dairy policy, total output would remain essentially unchanged due to the inelasticity of demand. Thus, substantially lower milk prices would be the outcome, and processors would be largely unaffected by BGH and a free market policy (Magrath and Tauer, 1986).

There is no consensus regarding the regional impact of adopting BGH in the United States. Michael Phillips, project director of the Office of Technology Assessment report on BGH impact (OTA, 1986), is quoted as saying that California-type corporate farms will dominate the dairy scene after BGH. Kalter counters this view on the grounds that north-eastern and mid-western states could remain competitive based on their comparative advantage of growing their own feed, while California must import part of it. For his part, Mix asserts that there are more intra-regional differences due to distinct managerial skills in farming, than inter-regional differences.

Therefore, his regional impact study of BGH predicts that Wisconsin will remain as the top state in milk production, but with much fewer farmers. Other beneficiaries, though, would be early adopters and those with greater managerial skills (Dupuis, 1985).

MILK PRODUCED AND CONSUMED BY THE POOR OF MEXICO

Contrary to what happens in the United States, Mexico has suffered from an under-supply of milk for many years. The production structure in Mexico is even more heterogeneous, with vast productivity differentials. Even the most highly modernised farms in Mexico have a productivity of only 60 per cent of farms in advanced countries (SARH, 1980).

Before we try to assess the potential impacts of BGH in Mexican dairy industry, let us briefly outline its current structure. After a period of expansion in 1960–77, the dairy industry has experienced a severe crisis in the past decade. This industry is made up of two broad types of operation: one which concentrates cattle in stables – the specialised sector – and makes up 12.7 per cent of the total herd, and another which has the double purpose of producing meat and milk – the non-specialised sector – making up 87.3 per cent of the total herd. The specialised sector produces 58 per cent of milk supply in Mexico, while the non-specialised sector produces only 42 per cent. These figures clearly reflect the vast productivity differences in the two sectors (SARH, 1980).

The destination of total milk production is as follows. Forty-five per cent is consumed directly, without pasteurisation, or is used for home-made milk derivatives; while 55 per cent flows to the processing industry. Of this total, only 40.5 per cent is pasteurised, 9.5 per cent is dehydrated and 50 per cent is used for industrial production of milk derivatives (SARH, 1980).

The specialised operations have followed the technical model of large American farms, but are only about 60 per cent as efficient, according to a study by the United Nations Food and Agriculture Organization (FAO). Moreover, modern farms do not create much employment: labour costs make up only 12.2 per cent of the total bill (SARH, 1980). With regard to foreign exchange, the cattle production complex is rather import intensive, given its technological model. While the agricultural sector as a whole accounts for 20.6 per cent of total Mexican imports, the cattle complex accounts for 50 per cent of agricultural imports (Arroyo and Waissbluth, 1988).

The processing end of the dairy industry is highly concentrated. Fifteen per cent of the firms process 67.5 per cent of pasteurised milk, absorbing 61.9 per cent of employees and generating a value added of 67.6 in this branch of the industry. In 1978, the five largest concerns alone produced 55 per cent of pasteurised milk in Mexico. The market of 'condensed' and 'evaporated' milk is controlled 100 per cent by two multinational enterprises: Nestlé and Carnation. With regard to milk derivatives, such as cream, cheese and butter, 2.2 per cent of the established concerns generate 50.1 per cent of total production (SARH, 1980).

Milk consumption is also rather concentrated, in so far as 85.5 per cent of available milk is consumed by 63 per cent of the total population; while 37 per cent of the population consumes only 14.5 per cent of available milk (including imports). Cattle feed supply is also highly concentrated, with three multinational enterprises exercising an oligopolistic control: Purina, Anderson Clayton and Hacienda. They control 47 per cent of the feed market in Mexico (SARH, 1980).

Horizontal integration among producers is very low. There are only a few cases where producers are organised to store raw materials, or purchase feed, machinery, equipment, etc. Backward and forward integration, however, is often promoted by the processing firms, basically by pasteurisers.

Mexican biotechnology research related to the dairy industry is virtually nil. A recent inventory of agricultural biotechnology found only two projects related to this field: one in the Ministry of Agriculture and Water Resources (SARH) on the establishment of an embryo bank; and the other in the Autonomous University of Nuevo León on BGH. Both projects were financed by the Federal Agency CONACYT (National Council for Science and Technology), in 1985 and 1986 (Arroyo and Waissbluth, 1988).

The Mexican dairy industry has experienced the worst crisis of its modern era in the past few years. Several of its leaders think that the crisis is so great and profound that it could have irreversible consequences for the industry if no firm and quick measures are taken to save it. In fact, production fell 28 per cent in the past three years (1986–1988), the national herd decreased from 950 000 heads in 1984 (in the specialised sector) to 700 000 in 1988, and much of the installed capacity is idle. Dairy leaders think that all of these problems can lead to a further decapitalisation of the industry, a loss of cattle grazing culture, unemployment in the countryside, and a massive drain of foreign exchange to import milk and other dairy products. On the nutritional end, FAO recommends a per capita daily allowance of 250 millilitres. Hence Mexico needs a total of 20 million litres daily for its 80 million people, but it is only producing 7.5 million in 1988, down from 10 million in 1984. The milk deficit for 1988 will be 12.5 million litres per day, some of which will have to be imported (Reyes and Neme, 1988).

What could happen if modern dairy farmers adopt BGH in their operations? It seems to us that, given the described situation of crisis in the dairy industry, two crucial things could happen. First, the new technology might become a factor in making the Mexican dairy industry more cost efficient, and thus be able to survive its current crisis. Second, this industry might be able substantially to increase milk supply locally, so that the drainage of foreign exchange through imports of dairy products may be reduced.

The problematic side of adopting BGH in Mexico is that it could have a further polarising effect in the current production structure of dairy farmers. This should be expected for those farmers who are immersed in the formal market and directly compete with modern operations. However, those who produce the 45 per cent of milk which is sold directly to consumers might be able to sustain their markets. Nevertheless, they will suffer the impact of decreased prices in the formal sector, should they emerge at some point. For many years milk prices have been controlled by the government, as part of the basic diet for Mexicans. However, dairy farmers are better organised than most agricultural producers, and they have exercised a strong influence in determining 'controlled' prices for milk. Their usual argument is that current prices fail to cover production costs. Perhaps the new technology might help them reduce costs so that prices become more stable or decrease *vis-à-vis* other prices.

Another problematic aspect of adopting BGH is that it would probably involve full dependence on importing the technology. However, the foreign exchange costs for the import of the technology would probably be rather small, compared to the costs of importing milk.

HIGH FRUCTOSE CORN SYRUP

Who benefits in the United States?

With the aid of new enzyme technology, new sweeteners have been produced which could profoundly change the sugar industry as we know it. The impact on employment in Third World countries, including Mexico, could be somber. This is all related to the international patterns of trade for sugar and its substitutes (Ahmed, 1988).

In the United States, for instance, between 1982 and 1987, 44 per cent sugar has been replaced by high fructose corn syrup (HFCS). Any vegetable with a high starch content may be used to make new sweeteners with modern enzyme technology. Like sugar, they also contain calories and are usually sweeter than sugar. Thus, we are not talking about dietetic substitutes for sugar, such as saccarin or asparthamus (one commercial name for the latter in the United States is nutrasweet). These substitutes also constitute a problem for the sugar industry, but this is a matter for a separate study.

The production of new sweeteners from high starch crops has taken place mostly after 1978. By 1985, world production was 6300 million tons of dry material, of which 4600 million tons, more than 66 per cent was concentrated in the United States (Arroyo and Arias, 1987).

Over 30 soft-drinks bottling companies in the United States (such as Coca-Cola, Pepsico and 7–Up) have shifted from the use of sugar to HFCS. As a direct result the sugar imports of the United States plunged from 4.6 to 2.5 million tons between 1978 and 1985. This change has lowered the cost of production processes, for the United States usually produces large quantities of corn in its Midwestern region. However, many Third World countries are suffering the consequences (Ahmed, 1988; Buttel and Barker, 1985).

In fact, such a change in the global pattern of trade relations has affected incomes and employment in the Third World. For instance, Caribbean income from sugar exports to the United States dropped from US$686 million in 1981 to US$250 million in 1985. On the other hand, the Philippines experienced a reduction of sugar exports to the United States from US$624 million in 1980 to US$246 million in 1984. One of the results has been the replacement of the sugar crop by others that are not as labour intensive. With this substitution the Philippines has lost half a million jobs (Ahmed, 1988).

The sugar industry in Mexico: Threats to employment

At this point, we should ask ourselves: what might happen if the new sweeteners are substituted for sugar from cane in Mexico. One of the dilemmas that would have to be faced, in the first place, is that Mexico does not even produce enough corn for human consumption, and corn is one of the most important components in the Mexican diet. Thus, there are no corn surpluses to produce HFCS.

Some features of the sugarcane production structure in Mexico are as follows. Its contribution to Mexican GNP decreased from 7 to 5 per cent in the 1970–80 period. Employment decreased in the same period from 4.1 to 2.6 per cent. Sugarcane accounts for 3 per cent of agricultural land in Mexico (Arroyo and Arias, 1987). With regard to land tenure, *ejidos* account for almost 69 per cent of land dedicated to sugarcane, while private farms, only 31 per cent. Most of this land, 60.5 per cent, is dependent on rain for its water supply, while only 39.5 per cent is irrigated. Yet, Mexico is the fourth largest producer of sugarcane, after Brazil, Cuba and India. By 1986, the Government controlled over 81 per cent of total sugar production. This proportion grew steadily since 1975, when the Government began to purchase many of the ailing sugar mills. The remaining mills are either privately or cooperatively controlled (Azúcar, S.A., 1987: various tables). The economically active population engaged in the sugar industry has evolved as shown in Table 7.1. Due to a price policy which kept sugarcane prices artificially low, and to consumption patterns encouraged by advertisement campaigns from the bottling companies, Mexicans consume more than 14.5 million litres of bottled beverages per day. In fact, Mexico is the second largest per capita consumer of such beverages in the world (after the United States).

Currently, industry in general processes 55 per cent of total sugar produced in Mexico and the rest is for household use. Of the grand total, though, just the bottling industry consumes 30 per cent. And from this, 75 per cent was controlled by two multinational enterprises in 1979: Coca-Cola and Pepsi-Cola. Therefore, much of the future control of the sugar industry in Mexico is in foreign hands.

In January 1988, the Mexican Government announced that it would sell 17 of the 56 sugar mills that it then controlled (out of a total of 66 mills). Pepsi-Cola soon expressed an interest in three of them: the most productive ones. Despite protests from sugarcane producers and sugar mill workers, who were bidding to purchase the mills, the Government has resolved to sell those mills to Pepsi-Cola. Another company comprising several soft-drink bottlers was established, but Pepsi-Cola owns a controlling interest in the new company.

If the sugar mills re-privatisation drive continues to transfer their ownership to multinational bottling enterprises, this could be a guarantee that they will not import HFCS from the United States. However, it will be inevitable that they restructure the plants, which are said to contain at least 30 per cent of redundant personnel. It remains to be seen how the new owners deal

TABLE 7.1 *Mexico: Economically active population dependent on the sugar industry*

Labour situation	1983	1984	1985	1986
Field Personnel	249537	254039	239798	240805
Plant Personnel	52940	56914	57180	52327
TOTAL	302477	310953	196978	293132

SOURCE: Azúcar, 1987.

with the problem of restructuring the plants, for sugar-producing regions are highly dependent on the sugar mills for income and employment.

One alternative for the new owners will be to try and develop other sources of employment locally for displaced workers in the sugar industry. Promising prospects exist in rationalising the use of by-products from sugarcane. The fact remains though, that an increasing number of crucial decisions for development are being transferred not only to private hands, but also to foreign hands. The question is thus whether this trend will be beneficial or detrimental for the majority of the people.

CONCLUDING COMMENTS

The dismal picture of biotechnology in Mexico leads one to believe that, should more Mexican entrepreneurs be interested in the emerging industry, they will at best seek joint ventures with American enterprises. At worse, multinational enterprises will attempt to transfer the required technology directly or through subsidiaries, with all the costly implications in terms of patents, royalties and higher forms of scientific, technological and, therefore, financial dependence of the Mexican people.

If decision makers in Mexico and Latin America have the will to intervene in this process, towards a national development of biotechnology, it will be necessary to act soon with firm measures of scientific and industrial policy. Of course, it would be illusory for Third World countries to attempt to develop the whole range of biotechnologies internally, with local scientific and financial resources. Nevertheless, there is a significant knowledge base in Mexico, as well as in Brazil, Cuba and Argentina. With this limited appropriate policy decisions and attempts at regional integration, such knowledge could be translated into indigenous technological developments. But the question would still remain whether biotechnology, as it is being developed today, is the right bandwagon for the Third World to jump onto, and whether the Third World has the choice to make this decision.

8 Biotechnology to Combat Malnutrition in Nigeria

G. U. Okereke

NUTRITION AND BIOTECHNOLOGY

The overall food demand in Nigeria is estimated to be growing at the rate of at least 3.5 per cent per annum due to the combined effects of population and income growth, while available data concerning the protein requirements of the Nigerian population show that the present production of food protein falls far short of demand (Idachaba, 1983). The intake of protein in Nigeria is 20–25 per cent below the satisfactory level (Nelson, 1972), while the average per capita total daily protein intake in Nigeria was 54 grams out of which only 8.3 grams were of animal origin (Norris, 1981). The value of animal protein intake is much below the 25 grams recommended out of a total protein requirement of between 65 and 75 grams. It is estimated that 1.68 million tonnes of beef would be needed annually to raise animal protein intake in Nigeria to 25 grams per capita per day (Olubajo, 1976). This is hardly possible in view of the present stationary or even declining cattle population, low productivity and small body size of indigenous livestock.

In recent years the expansion of meat production in Nigeria has been mainly in poultry production. The poultry industry has a greater contribution to make in filling this wide gap of protein requirement in the Nigerian diet. This is because poultry has a relatively shorter 'gestation' period than the other classes of livestock. Also poultry products are more accessible and acceptable to Nigerians of different income brackets. Nigeria in recent years has entered a period of increasing food shortage. Thus protein as an animal feed component will indeed become increasingly scarce and expensive. Abnormal climatic conditions are largely responsible for the sharp fall in protein availability. It was the abnormal protein supply conditions prevailing in the mid-1960s that led to rapid development of the technology of Single Cell Protein (SCP) production (Norris, 1981). SCP is a rich source of protein which can be used as a supplement to animal feed for chickens (broilers and laying hens), calves, sheep and

127

TABLE 8.1   *Compositions of representative single-cell proteins*

SCP	Substrate	Composition (wt%)				
		Crude Protein	True Protein	Lysine	Methio- nine	Fats
Bacteria	Methanol	80	65	5.8	2.2	8
	Methane	60	50	4.3	3.0	10
Yeast	n-paraffin	60	53	7.4	1.8	9
	Gas oil	69	60	7.8	1.6	2
	Ethanol	54	45	6.7	1.5	6
Fungi	Carbohydrate	35–40	30–40	6.5	1.5	5
Algae	CO_2	45–60	40–50	4.6	1.4	5
Fish meal		60–65	50–60	7.0	2.6	5–10
Soy meal		45–50	40–45	6.5	1.4	1.5

SOURCE: Schwarts and Leathen, 1978.

farmed fish. When compared to conventional feeds such as soyabean meal, SCP has a significantly higher crude protein content (Table 8.1) and better rate of metabolisable energy. On a net protein value basis, it is estimated that a unit of SCP is equivalent to 2.15 to 2.26 units of soyabean meal (Senez, 1985).

SCP is now widely accepted for use in a variety of animal and some human foods (Scott, 1983; Hamdan, 1983). SCP production would reduce Nigeria's heavy dependence on imports of animal feed protein without massive developments in agriculture. It has been stated that the solution to balancing the supply of and demand for protein is not just the import of more food, whether SCP or other food products by protein-short countries, but in the development of technologies in these countries so that they will be able to produce more food (Norrel *et al.*, 1983).

The introduction of the SCP technology into Nigeria, whose economy is primarily agricultural, would have many interrelated social and economic implications. As with any new industrial project, SCP has to demonstrate that it is commercially viable. This viability will depend on demand, production costs and competitiveness with conventional sources of protein (Senez, 1985).

The British Petroleum Company (BP), which was the first to enter into commercial production of microbial food and feed, in 1974 shut down most of its plants in various countries because of high crude oil prices, low soya and fish meal prices and various health problems caused by eating food produced on hydrocarbons (Chahal, 1987).

The American corporation, Gulf, also entered the field of microbial feed production, but abandoned its plans for commercialisation several years ago. Their decision was based on economic considerations (Laskin, 1977). Therefore, a study has to be carried out on the economic viability of SCP, including such considerations as its capacity to substitute for soyabean, availability of raw materials for the production of SCP and the size of its market.

OBJECTIVES

The major objectives of the study were:

(a) to assess the economic and technical potential of biotechnology applications in the animal feed industry;
(b) to examine biotechnology's contribution to combating protein malnutrition and facilitating import substitution;
(c) to examine the socioeconomic and employment implications of the use of biotechnology; and
(d) to review policies and measures to promote the diffusion of biotechnology.

Methodology

The methodology for the study involved informal interviews of poultry feed millers and users on questions pertaining to feed and poultry production. The method also involved the use of questionnaires which were designed for collection of data from feed millers, and poultry farmers who are users of the feed of which SCP is a component. Additional information was collected from secondary sources such as journals, bulletins and Annual Report of Poultry Association of Nigeria, and discussions with Ministry of Agriculture.

SINGLE CELL PROTEIN TECHNOLOGY

Protein malnutrition in the Third World is usually far more severe than that of other foods. Micro-organisms could help meet the world's protein deficiency. Of course, SCP cannot completely replace the need to harness proteins from plants such as oil beans or from animals such as fish. However, the limitations of conventional sources of proteins were recognised. These include: (a) possible crop failure due to unfavourable climatic conditions in the case of plants; (b) the need to allow a time lapse for the replenishment of stock in the case of fish; and (c) the limited land available for farming in the case of plant production.

On the other hand, the production of SCP had a number of attractive features: (a) it was not subject to the vicissitudes of the weather and could therefore be produced all year round. It does not require use of soil, neither does it require large space; and (b) micro-organisms have more rapid growth than plants and animals. The biggest economic attraction is the ability of these micro-organisms to utilise waste products such as petroleum by-products as the raw material substrate. In terms of nutrient content, it has a high protein content, between 60 and 70 per cent (Table 8.1). The protein content is much higher than that found in plant and even animal sources. A switch to SCP is itself not entirely free of constraints. One of the most obvious is that many developing countries, where protein malnutrition actually exists, lack the expertise and/or the financial resources to develop the highly capital-intensive fermentation industries involved. But this shortcoming can be overcome by the use of improvised fermentors and recovery methods which do not require sophisticated equipment. Other criticisms of SCP are that micro-organisms contain high levels of RNA and that its consumption could lead to uric acid accumulation, kidney stone formation and gout.

MICRO-ORGANISMS USED IN SCP PRODUCTION

A list of selected micro-organisms currently used in SCP production is given in Table 8.2. Organisms to be used in SCP production should have the following properties:
(a) Absence of pathogenicity and toxicity: it is obvious that the large-scale cultivation of organisms which are pathogenic to animals or

TABLE 8.2   *Status of micro-organisms for SCP*

Name	Status
Saccharomyces	Food, accepted if grown on sugars (including whey) or ethanol (even if petrochemical based); feed, general acceptance.
Candida, Hansenula	Wide acceptance as feed.
Paecilomyces	Accepted as feed.
Pseudomonas	Yet to be approved as feed.
Fusarium	Awaiting FAO approval as feed.
Aspergillus	Test evaluation completed.
Kluyveromyces	Feed.

SOURCE: Adapted from Kristiansen and Bulock.

plants could pose a great threat to health and therefore should be avoided. The organism should not contain or produce toxic or carcinogenic materials.

(b) Protein quality and content: the amount of protein in the organism should not only be high but should contain as much as possible of the amino-acids required by humans.

(c) Digestibility and organoleptic qualities: the organism should not only be digestible, but it should possess acceptable taste and aroma.

(d) Growth rate: it must grow rapidly in a cheap, easily available medium.

(e) Adaptability to unusual environmental conditions: in order to eliminate contaminants and hence reduce the cost of production, environmental conditions which are antagonistic to possible contaminants are often advantageous. Thus, strains which grow at low pH conditions or at high temperatures are often chosen.

SUBSTRATE AVAILABILITY

The choice of substrate for microbial protein production will depend on the availability of suitable substrate in a particular country. For example, hydrocarbon may still be the main and cheapest carbon source for SCP production in the Middle East where other carbon sources (non hydrocarbons) are in limited supply (Hamer, 1977). Crop residues (straw) may be the best choice for India, Bangladesh, Mexico, Pakistan and Indonesia; and sugarcane bagasse for Brazil, Cuba and India (Ishaque and Chahal, 1987). On the other hand, forest residues may be the best choice for Canada and other countries (Rawat and Neutyal, 1987).

The possible substrates for SCP production can be classified under three major groups: (a) energy source materials – natural gas, n-alkanes, gas-oil, methanol, ethanol and acetic acid; (b) renewable source materials – starch, sugar and cellulose and (c) waste material. Figure 8.1 shows details of the different waste materials and micro-organisms potentially available for SCP production.

Nigeria could develop SCP technology that uses labour intensive fermentor methods and other equipment fashioned from relatively cheap materials. The importance of the Nigerian petroleum and natural gas resources encourages the development of SCP technology as well as mobilisation of human resources for the working of these resources at a pace consistent with Nigerian economic development. With proven reserves of about 19 million barrels of oil (and about 115 billion barrels of potential oil reserves) Nigeria controls about 34 per cent of Africa's total reserves

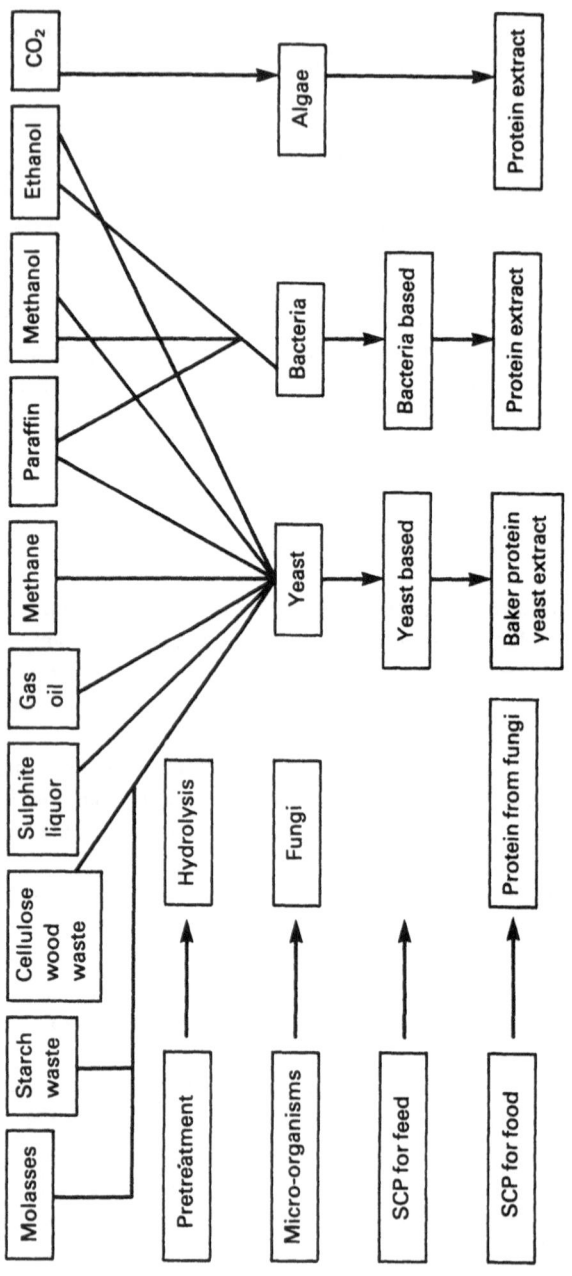

Figure 8.1   Substrates for production of SCP from classes of micro-organisms

and about 3.2 per cent of the entire world's proven reserve (Egbogah and Aikhionbare, 1980). It has also been estimated that Nigeria's gas potential is enormous, probably exceeded only by those of the USSR, the Islamic Republic of Iran and Algeria (Egbogah, 1978).

ECONOMIC IMPLICATIONS OF SCP TECHNOLOGY

The importance or utility of any process depends upon its economics. It must be profitable to be commercially developed and industrially widespread. The future of SCP as a protein source for feed is involved not only in current supply and demand but also in the great protein needs of the future.

One factor that influences the demand for animal protein is the growth in per capita income of the consuming nation. The oil wealth in Nigeria and the consequent economic development it engendered tended to increase the purchasing power of her population during the oil boom period.

Estimates by experts in the Federal Ministry of Agriculture (Table 8.3) show that the income elasticity of demand for beef is 0.9, while that of poultry is 1.2 and for eggs it is 1.0. The same source also showed that the demand for proteinous food like beef, goat meat and poultry, was on the increase and that by 1985, per capita consumption was projected to double what it was in 1968.

In this section the projected demand for poultry products will be considered since factors affecting them would indirectly affect the demand for poultry feed stock, thereby also affecting SCP production (a proxy for

TABLE 8.3 *Projection of demand for protein foodstuffs in Nigeria: 1968–1985*

Meat type	Income elasticity	Kilograms per capita per annum			
		1968	1975	1980	1985
Beef	0.9	3.338	4.329	4.795	5.336
Goat meat	0.9	1.840	2.386	2.643	2.941
Mutton and lamb	0.9	0.422	0.548	0.609	0.675
Poultry	1.2	0.839	1.217	1.406	1.616
Pork	0.6	0.482	0.522	0.577	0.612
Offal	0.6	6.454	7.385	7.728	8.197
Bush meat	0.8	3.704	4.416	4.809	5.279
Fish	1.0	13.379	18.044	20.275	22.806
Egg	1.0	0.839	1.132	1.271	1.432

TABLE 8.4 *Poultry, meat and egg demand in Nigeria*
Projections to 1990

Product	Actual Estimates 1977/78	Projections (Million kg) 1979/80	1984/85	1989/90	Average Annual Growth Rate
	(million kg)				(1978–90)
Poultry, meat	88.98	99.21	119.23	143.32	4.86
Poultry, egg	247.34	275.72	331.42	398.35	4.86

feed poultry). The demand for poultry feed is a derived demand, in the sense that only poultry farmers would need the feed for their chickens. The demand for feed (SCP) would largely depend on the nature of the poultry industry and willingness of people to go into poultry and other livestock production. Table 8.4 shows the projected potential demand for poultry and egg and Table 8.5 shows the projected supply up to 1990.

These projections glaringly show that the demand in poultry and poultry products is greater than the projected supply for the period covered. Hence the potential demand for poultry feed is established. The oil glut of the 1980s and acute shortage of foreign exchange in 1983, which led to the imposition of import restrictions, dealt a severe blow to the poultry industry. The price of maize rose from N250.00 per tonne in 1983 to over N1000.00 per tonne in 1984, but because so many farmers, including feed millers, later went into direct production, the price of maize had to drop up to N500.00 per tonne in 1987. Apart from maize, prices of every other ingredient in poultry feeds – especially groundnut cake, soyabean, meal,

TABLE 8.5 *Poultry, meat and egg supply in Nigeria*
Projections to 1990

Product	Actual Estimates 1977/78 (million kg)	Projections (Million kg) 1979/80	1984/85	1989/90	Average Annual Growth Rate (1978–1990)
Poultry, meat	67.72	71.64	79.18	87.70	2.42
Poultry, egg	254.16	266.44	289.75	316.62	2.00

TABLE 8.6   *Average price increases of poultry feeds from 1983 to 1987*

Type of feed (concentrate)	Price of 25kg bag		Per cent increase
	1983	1987	
Broiler	15	58	286
Grower	11	33	200
Chick	9	60	567
Layers	18	46	156
Breeders	12	52	333
Broiler starter mash	12	28	133
Growers mash	9	20	122
Chick mash	11	25	127
Layers mash	11	22	100
Broiler finisher	11	25	127

Source: Field survey, 1988.

wheat offals, vitamin pre-mixes, and so on, have increased astronomically. This is so because of the excessive import of vegetable oil; the oil millers were forced to charge the value of their unsaleable locally-produced vegetable oil on to groundnut cake and soyameal, which is the cheapest source of protein available in this country. Also, the cost of poultry feeds (Table 8.6) has become so excessive that the poultry farmers can no longer purchase feeds to produce eggs and chickens at a price the consumers can afford.

The trend in the prices of these raw materials over the years reflected the trend in the changes in the various national fiscal policies of the Federal Government of Nigeria. This upward trend in prices of the various raw materials for feed production directly affected the prices of the different feeds as shown in Table 8.6. These prices drastically diminished the profit expectation of poultry projects.

TABLE 8.7   *Preferred type of protein*

Preferred order of	Soya or other vegetable proteins		SCP (bacteria yeast)		No preference	
	Number	%	Number	%	Number	%
First	20	77	6	23	7	4
Second	6	23	19	73	0	0
Thirdl	0	0	1	4	25	96

SOURCE: Field Survey, 1988.

As a result of high feed prices and other management problems most firms operated their plants below 50 per cent of their capacity while others have completely closed down. This led to high average cost of production and a price which was too high to allow it to compete effectively in the marketplace. The poultry feed supply problem is aggravated by the poor distribution of infrastructure. Every poultry producer wants to be assured of a steady, unfailing supply of fresh nutritive and palatable feed.

Some of the feed millers, who attempt to maintain high quality, run into trouble when they are unable to obtain some of the essential ingredients, such as soyabean and fish meal. Evidence shows that poultry producers have been handicapped by the insufficient and irregular supplies of poultry feed. Moreover, as the cost of feeds in poultry production amounts to about 65 per cent of the total costs, which means that the failure of the feed industry implies the failure of the poultry industry.

The preferred type of protein was also considered: nearly 77 per cent of the respondents chose soyabeans first, but 73 per cent chose SCP second (Table 8.7). The fact that most respondents preferred soyabean was not unexpected. To date, no SCP is available in commercial quantities in Nigeria, and since the processor is familiar with soyabean products, he is more likely to respond to them. These results suggest that the SCP producer must educate the processor about the relative value of SCP as a food ingredient. If SCP is to be accepted with a minimum of delay (independent of the safety question) the ways in which it will be used, its economic benefits, as well as the potential problems that might arise if it is used improperly, must be thoroughly considered. To achieve this the reactions of the feed millers and the possible consumers of products containing SCP must be determined.

Most respondents declared themselves ready to incorporate the SCP in their feeds, as long as the product is nutritious and safe. This assurance can be given on the basis that large-scale feeding in some industrialised countries has demonstrated that, with care and special processing, problems of acceptance and toxicity can be overcome.

In a survey of feed millers in Nigeria it was found that, *ceteris paribus*, the substitution of SCP for plant protein sources would vary depending on the price of SCP relative to those of traditional sources. For example, if SCP is sold at the same price as the traditional protein sources, 69.2 per cent of the millers would effect at least, 41 per cent substitution (Table 8.8). The single highest category was the 41–60 per cent substitution indicated by 38.5 per cent of the feed millers.

However, where the SCP is about 10 per cent more expensive (even if more available), the substitution changed (Table 8.9). The single highest

TABLE 8.8   *Substitution of SCP for traditional sources of proteins when the costs are the same*

Amount of substitution	Respondents		
%	No.	%	Cumulative %
100	3	11.5	11.5
61 – 99	5	19.2	30.7
41 – 60	10	38.5	69.2
21 – 40	3	11.5	80.7
11 – 20	5	19.2	99.9
TOTAL	26	99.9	

SOURCE: Field Survey, 1988.

TABLE 8.9   *Substitution of SCP for traditional sources of proteins when SCP is 10 per cent more expensive*

Amount of substitution	Respondents		
%	No.	%	Cumulative %
100	2	7.7	7.7
61 – 99	0	0.0	7.7
41 – 60	4	15.4	23.1
21 – 40	11	42.3	65.4
11 – 20	6	23.1	88.5
5 – 10	3	11.5	100.0
TOTAL	26	100.0	

SOURCE: Field Survey, 1988.

TABLE 8.10   *Substitution of SCP for traditional protein sources when the SCP costs 10 per cent less*

Amount of substitution	Respondents		
%	No.	%	Cumulative%
100	13	54.2	54.2
61 – 99	4	16.7	70.9
41 – 60	3	12.5	83.4
21 – 40	3	12.5	95.9
11 – 20	0	0.0	95.9
5 – 10	1	4.2	100.1
TOTAL	24	100.1	

SOURCE: Field Survey, 1988.

TABLE 8.11 *Substitution of SCP for traditional protein sources when the SCP costs the same as the traditional sources, but is more available*

Amount of substitution %	Respondents		
	No.	%	Cumulative%
100	9	37.5	37.5
61 – 99	3	12.5	50.0
41 – 60	8	33.3	83.3
21 – 40	3	12.5	95.8
11 – 20	0	0.0	95.8
5 – 10	0	0.0	95.8
TOTAL	24	100.0	

SOURCE: Field Survey, 1988.

category was 21–40 per cent indicated by 42.3 per cent of the respondents. Here only 23.1 per cent of the respondent feed millers would replace 40 per cent or more of their traditional protein sources with SCP.

If the price of the SCP were, however, 10 per cent less than that of the conventional sources, then 54.2 per cent of the feed millers would go for 100 per cent substitution (Table 8.10). In fact, 83.4 per cent of the millers would go for, at least, 41 per cent substitution. Thus, the price of the SCP relative to that of the traditional protein sources is a critical factor in the adoption of SCP by livestock feed millers.

Another important factor is the availability of SCP. As Table 8.11 shows, even when SCP and traditional protein sources cost the same, more feed millers would use SCP instead of the traditional protein sources (compare Table 8.9 with Table 8.11). The Nigerian feed millers are also concerned about the safety of SCP, its nutritive value, cost and flavour.

Survey results show that their first concern was the nutritive value of SCP in terms of egg yield of layers, weight gain of broilers in particular, and chickens in general. The second in importance was the safety of SCP in terms of the health of the chickens. The third was cost which is closely associated with profitability. With feed prices rising slower than feed input costs, the profit margin was being squeezed. Flavour was also important to the feed millers, although this factor was rated fourth, behind the three factors already discussed. Nearly 70 per cent called for public education about SCP to enhance its adoption by both the millers and the chicken producers. SCP can be used either for animals or for human beings as long as impurities or compounds that might be considered as health hazards are removed (Mariscal and Viniegra-Gonzatez, 1977). This requirement is important since (as already indicated) *Candida sp* has approximately 8 per cent of RNA which causes metabolic disturbance when taken without any prior treatment (Edozien and Scrimshaw, 1970). Most of the technical problems of SCP production have been solved in such places as Kuwait as well as other developed countries. Extensive nutritional and toxicological studies conducted on the SCP product from both pilot and full-scale plants have proved it to be safe, highly nutritious and of good quality (Hamdan, 1983).

BENEFICIARIES OF SCP TECHNOLOGY

The primary beneficiaries of the introduction of SCP into Nigeria include poultry farmers, consumers, feed millers and government poultry farmers, who will benefit from consistent and available plant protein sources, thereby indirectly implying constant supply of high-quality feeds. As mentioned earlier, many poultry farmers went out of business because of the high cost of poultry feeds and other inputs used in production: diminishing returns made it difficult for many farmers to break even. Also, the layers did not start laying until after seven months instead of the usual five months. Even when they started to lay, the egg production efficiency was less than 50 per cent. This may be attributed to the use of low-quality feeds.

Feed millers will benefit from the availability and consistency of the SCP. With the ban on the import of grains and other feed ingredients, many feed millers went out of business owing to lack of raw materials. This situation will be removed when the materials for the production of SCP are based on raw materials available locally. Furthermore, the feeds produced in Nigeria were inconsistent in terms of nutrient composition. When protein components became scarce, the millers simply increased the carbohydrate proportion, and the unscrupulous ones were even known to include sawdust. In addition, even when the protein components were available, the percentage protein content varied from consignment to consignment. With the consistency expected from SCP, such problems should not arise.

Consumers are expected to benefit from reduced costs arising from increased productivity of layers, broilers and other chicken enterprises. For example, improved quality feeds will reduce the age at point-of-lay from about eight months (with bad feeds) to five months. Also, the egg laying efficiency could increase to 80 per cent or more with good management. The same positive increases are expected in broiler production and other chicken enterprises. Consumers will also benefit from higher-quality poultry and poultry products, as well as from the increased availability of these products.

The Government will benefit indirectly through the acquisition of SCP technology by Nigerians, reduction in the costs of poultry and poultry products, increased activity in the poultry sub-sector (employment, income, investment and profit), and a healthier population. The employment and income gains will be derived from forward and backward linkages in the system. The production of SCP will involve inputs locally generated. The employment and income generating prospects of the introduction of the SCP technology into the Nigerian economy can be traced from the input supply stage to the poultry consumption stage (the final product). The largest employers of labour are expected to be the poultry producing and processing sub-sectors through expanded activities. Of course, in the final analysis, its impact on employment and income will depend on the extent of its adoption. In the poultry sector, it has been found that labour constitutes 10 per cent of the total cost of production (Akinwumi, 1979).

Increasing SCP for animal feeds would therefore improve human nutrition rather quickly, by taking vegetable sources of protein out of the human/animal competition and making them available for human consumption in Nigeria. In the face of uncertain crop yields and the ever increasing Nigerian population and its demand for poultry products, the feed millers will be attracted to SCP because it is a product of consistent composition with good biological value, which can be produced under highly controlled conditions and is independent of soil and climate. Because of the proven enormous potential of the SCP products, as shown by reports from the developed countries, SCP production, if practised in Nigeria, will supply the ever dwindling feed industry with good-quality proteins that will not only fill the deficit created by escalation of protein prices, especially soyabean, during the last few years but also create more jobs in Nigeria. It will not lead to unemployment because the producers of vegetable feeds will now produce for human consumption. *This will reduce the price of soybean as there will be no competition for it from the point of view of the feed millers.*

CONCLUDING REMARKS

It is expected that the market need for SCP will continue to rise as the demand for poultry products, which are an animal protein source for the majority of the urban and rural population in Nigeria, also continues to rise. To allocate funds to import poultry products would constitute a severe drain on the foreign exchange reserve of the country. In realisation of this the Federal Government of Nigeria has promulgated a decree banning such food and feed imports into the country. Hence the need to look into alternative sources of protein for compounding feed for the poultry industry is very obvious. In addition, sustained high levels of price for both fish meal and soyabean meal point to the likelihood of a continuing market for SCP in Nigeria. SCP will complement vegetable proteins in order to help improve the dwindling feed and poultry production in Nigeria.

There are several reasons why SCP production provides a potentially valuable alternative protein production system in Nigeria. The crude protein content of SCP ranges from 60 per cent upwards, compared with 45 per cent for soya meal and an average of 65 per cent for fishmeal (Selignam, 1976). Nigeria, as one of the world's leading producers of the raw materials for SCP production, stands a better chance in terms of ready availability of one of the major substrates (natural gas) which at present is being flared to the tune of 400000 barrels equivalent of petroleum, roughly $15000000 every day (Osakwe, 1981). Gas which so far has not yet been efficiently exploited can be recovered and utilised in the SCP production especially with the current import ban on some food materials in Nigeria.

SCP production may constitute a very attractive alternative for those countries with hydrocarbon-based resources *that have a weak agriculture* and, most importantly, the *desire* to be more self-sufficient in animal production (Hamdan, 1983).

Nigeria has these strengths and therefore SCP technology is likely to be a successful and adoptable technology in the country. The SCP research in Nigeria will enable the development of new techniques for the growth of micro-organisms that effectively utilise the abundant by-products of petroleum and other waste materials. The research and development of SCP will, therefore, be targeted towards the specific end of a technically and commercially successful process. This is with the ultimate aim of producing high protein feed-stuff components, in sufficient quantity to complement the use of soya meal and fish meals in the diets of intensively farmed animals such as poultry. The R&D strategy for the development of SCP in developing countries should not be based on direct economic returns in the short run but rather on long-term planning and should recognise the need to achieve self-sufficiency in food supplies to foster independence (Hamdan, 1986).

As SCP research and development has not been widely carried out in

Nigeria we recommend that part of the R & D programme should follow the following steps:

(a) Isolation of SCP bacteria and substrate.
(b) Studies on the physiology and growth characteristics of promising cultures.
(c) Pilot production facilities for scale-up and technology development.
(d) Studies on cell recovery and drying processes.
(e) Nutritional and toxicological studies.

The development of SCP production could be suggested as an additional and/or complementary route to food conversion by domestic animals, but not necessarily as the exclusive way of solving the problem of food demand and consumption (Mariscal and Viniegra-Gonzatez, 1977). This is in agreement with what has been established by the Protein Advisory Group of the United Nations, that 'the protein from micro-organisms offers the best hope for a new source of major protein, independent from agriculture' (Protein Advisory Group of the United Nations System, 1974).

The benefits of the SCP should be seen in terms of the socioeconomic needs of the poultry farmer, feed miller, government and their family, for example from the point of view of their nutrition, food security, employment and income.

In a developing society such as in Nigeria, the commitment to use SCP technology for compounding poultry feed must be accompanied by an intensive activity related to the generation of technical manpower through appropriate education policies, setting up of R & D institutions, formulation of policies for the generation, transfer and adaptation of the SCP technology. Data on Tables 8.8 to 8.11 show conclusively that the relative prices of SCP and conventional sources of feed can be used as a major policy instrument for the diffusion of SCP technology by government.

One of the possible approaches to commercialising SCP technology could be through government assistance to entrepreneurs to bear the risk involved in commercialisation of the new technology. The Government could set up a technology production company in collaboration with private firms to manufacture and market such technology on a commercial basis. It may later sell out its shares to private companies and entrepreneurs, after the risk phase has passed. In carrying out such a programme the Government might enact various laws aimed at protecting the market for such technology or product.

If Nigeria directs her interests towards developing the SCP technology and product for poultry feed production, the much orchestrated and anticipated food crisis and protein malnutrition, as a result of the rapid population growth rate, will be reduced. Nigeria should move forward to embrace industrial production of this new source of food. This is because the wealth of a country depends on the production of goods and services through the co-ordinated use of the available supplies of human skills, capital and available raw materials.

Part IV
The External Threat

9 Biotechnology Trends: A Threat to Philippine Agriculture?

Saturnina C. Halos

INTRODUCTION

One can argue that Philippine agriculture is more a victim rather than a beneficiary of technological breakthroughs and new scientific findings. Its once lucrative sugar industry has been brought to death's door by the advent of artificial sweeteners and the high fructose corn syrup. The world market for coconut products has not expanded with population increases as expected because of improvements in palm oil and soybean technologies. The finding that aflatoxin is carcinogenic and is contained in Philippine copra meal at levels above EEC standards could close the European market for this coconut product. On the other hand, unless government makes credit available, the modern rice and corn production technologies would remain beyond the reach of the majority of the Philippine farmers who are mostly poor and cash-starved.

There are several reviews that assess how these developments would affect Third World agriculture (Junne, 1990; Buttel et al., 1985). Ahmed (1988) has argued that the bio-revolution has both pro-poor and anti-poor potentials, and the apparent trend is more anti-poor.

This chapter analyses trends in biotechnology research and development, both national and external in relation to Philippine agriculture. The following questions are being asked: Which of the current R&D activities affect the world market for Philippine major export crops, such as sugar, coconut and fruit crops? How do these research results affect the production of these crops and subsequently the small farmers? Are existing extension/technology transfer structures suitable for the dissemination of appropriate biotechnologies to small farmers? Are existing trends, programmes and structures in local R&D sufficient to propel Philippine agriculture into the fold of the bio-revolution?

Basically, the chapter examines the employment potential, distributional implications and nutritional dimension of the locally-developed biotechnologies. In this respect the study covers both new biotechnologies and the traditional ones for which the socioeconomic impact has not been adequately explored.

This chapter is divided into five parts. The first part describes briefly Philippine agriculture and agricultural plans and programmes; the second discusses international biotechnological developments that would be likely to affect Philippine agriculture; the third part describes biotechnology activities in the Philippines; the fourth part discusses how some agricultural biotechnologies being developed in the country may affect the Filipino farmer; and the fifth discusses specific action the Philippines can take to benefit most from the developments in biotechnology.

Philippine agriculture: plans and programmes

The Philippines has about 12 million hectares of arable land, more than nine million of which are devoted to rice, corn and coconut, and the rest to rootcrops, vegetables, fruits and other crops. The principal export commodities used to be semi-processed or agricultural products derived from coconut and sugarcane which in recent years have been replaced by manufactured goods.

Agriculture has always been the biggest employer of labour. The National Census and Statistics Office data show that as of the first quarter of 1985, agriculture employed 9.293 million or 46.5 per cent of the total labour force, a slightly lower figure than the previous range of 50.4 per cent recorded from 1956–75. Women comprise more than 20 per cent of this labour force.

Farm families grow various crops and engage in both farm and non-farm income-generating activities. Despite these, poverty stalks the agricultural sector. The average farm family income per month of about $63.57 (indicative figure) is below the poverty line of $83.71 per month and well below the rural average family income of $89.76 per month (Department of Agriculture, 1987). This has been attributed to low productivity which in large part could be due to poor crop production. Yields per unit area are very low compared to those obtained in other countries or to expected yields. Poor yields are associated, among other things, with unfavourable climate, poor soil and poor technology.

The stated goal of the Department of Agriculture is to attain equity and economic recovery. The primary objective is to increase farm family income from the present indicative level of $63.57 per month to at least $95.24 per month by 1992. The other related objectives are to ensure food security, increase and stabilise earnings from agricultural exports and reduce dependence on products where the comparative advantage is threatened. The overall strategy is to provide a policy environment that will make farming profitable and improve the delivery of services to farmers. It has envisaged a private sector led development process with bottom-up rather than top-down planning, with programmes that are area-specific and community-based rather than commodity-specific and national in scope, and that are income-oriented rather than production-oriented.

The unequal distribution of land is recognised as a disincentive to farmers and landless workers; hence, assistance will be provided to beneficiaries of the Comprehensive Agrarian Reform Program. Programmes to assist the poorest of the poor, such as the upland farmers, cultural communities and subsistence fishermen, are being implemented. On the other hand, the rapid depletion of natural resources reduces the productivity and endangers the long-term sustainability of agriculture.

Crop diversification is being promoted as a solution to the vulnerability of traditional exports (sugar, coffee and coconut) to world commodity markets, to the inadequate supply of some basic food items like beef and dairy products and to dependence on imported raw material inputs for agro-based industries.

The need to develop and disseminate technologies to improve agricultural production through research and extension is recognised. The following are the relevant issues: the inadequate funding of agricultural research which is only 0.2 per cent of the agricultural gross value added in 1982; research priorities not aligned with farmer's needs; the fragmentation and duplication of research activities; and the slow pace of technology transfer due to weak linkage between research and extension. Several policies and programmes have also been formulated to promote/expand the market for agricultural products which includes trade liberalisation, market development and infrastructure support.

BIOTECHNOLOGY DEVELOPMENTS ABROAD: A THREAT TO THE POOR

This section presents and cites specific effects on the Philippine agriculture of general biotechnology trends, namely the increasing privatisation of biotechnology, the search for non-traditional uses of crops, the consequences of biotechnology developments arising from health fads, and new forms of production through biotechnology.

The increasing privatisation of biotechnology, with the consequences of bringing knowledge and new technologies including new plant varieties and organisms under the control of the industrialised countries, has been widely recognised. Consequently, research is oriented to meet the needs of an affluent market such as the high value-low volume products in the pharmaceutical industry. Similarly, agricultural technology packages being developed by large multinational enterprises consist of improved seeds plus corresponding pesticides such as herbicide-resistant corn. Moreover, the herbicide-resistant corn is not likely to immediately benefit the majority of corn farmers in the Philippines just as hybrid seeds had not. Corn hybrid seed adoption in the Philippines is a meagre 4 per cent despite several subsidised corn production programmes and the fact that several hybrids have been approved by the Philippine Seed Board since 1961 (Mendoza, 1985). This low adoption rate could be attributed to the failure of corn production programmes to reach out and cater to the needs of the small corn farmers who produce 75 per cent of the total corn harvest in the country. Small corn farmers grow corn primarily for subsistence and prefer the open pollinated, white varieties which are more suitable for food and can produce the seeds for the next season's crop without substantial loss in yield. In the past, corn production programmes have been focused primarily on yellow (feed) hybrids, the seeds of which must be bought for every season's crop. It is likely that corn technology packages being developed will be geared towards the feed market, since the market for feed corn world-wide and in the Philippines itself is substantial.

On the other hand, these technology packages represent another cost-reducing tool for the farmers of the United States and other industrialsed countries, thereby increasing their competitive edge and reducing world corn prices. Without new and more appropriate technologies the Filipino corn farmer may not even be given a chance to compete at all even in his own market, especially if corn import is liberalised.

Thus, it is necessary to maintain corn-breeding programmes in public institutions in developing countries like the Philippines. It is through these programmes that benefits in biotechnology can be achieved without depending on the genetically-manipulated parentals developed elsewhere. This brings to the fore an issue of the current trend of patenting such organisms. Although a patent does not keep a technology from being adopted, the question in this case is whether it is fair for the Philippine public to pay for improvements made on one or two properties of the corn plant improved by biotechnology, when the public of industrialised countries did not pay for the domestication of corn and other crops originally obtained from the Third World but now grown in the West.

The continuing search for non-traditional uses of crops through genetic manipulation and bio-conversion would most likely affect the market of industrial crops such as sugar and coconut grown in the Philippines.

Cassava, a root-crop that traditionally supplies starch for industrial and pharmaceutical use, is being considered as a source of high fructose syrup (HFS), a sweetener substitute for sugar. Thailand and Singapore are developing biotechnologies for the conversion of cassava starch into high fructose syrup, making cassava competitive with sugarcane. High fructose syrup from corn has eased the Philippine sugar out of a major portion of the United States sugar market. Undoubtedly, HFS from cassava would force the Philippine sugar to seek other markets as well. Otherwise, the industry may suffer another slump similar to that seen in 1984–86 when production, hectarage and employment dropped significantly. The result has been widespread poverty, malnutrition and increase in insurgency. The hardest hit was the rural labour market whose size is significant. It was created by, and is dependent on, the sugar industry.

Of the new production systems that are currently being developed, the possible commercial production through fermentation of thaumatin, a plant protein that is a possible sweetener and sugar-substitute and whose gene has already been cloned into a microbe could further hurt the Philippine sugar industry. Thaumatin, aside from being 200 times sweeter than sugarcane, has the added advantage of being low-calorie and possibly highly nutritious.

The rapid pace of research on improving yields of various seed oil crops will not only widen the price gap between coconut and other seed oils but could also create a glut in the seed oil market. This together with efforts to use enzymatic conversions of plant oils to produce structured lipids or tailored fats could in the future eventually ease the coconut oil out of the market. Only 2 per cent of the oils and fats market is supplied by coconut, and the Philippines supplies about 80 per cent of that coconut (Sangalang, 1985).

Structured lipids or tailored fats would allow the industry to produce oil formulations that simulate the desirable properties without the concomitant unwanted properties of various seed oils and natural fats. These simulated products could, therefore, be highly competitive with coconut and could displace it from its share of the seed oils market as is elaborated in Chapter 10.

Fifteen million Filipinos are dependent on the coconut industry for livelihood and from the 1982–83 data, the majority are poorer than the rest of the farming sector (Padolina, 1985). The industry has long been suffering from the unpredictability of the world market and a major technological advance such as simulated oils and fats could restructure the market leaving a very low share for coconut oils. Unfortunately, the ability of coconut to compete is limited by its long gestation period, so that interventions to improve farm productivity are not immediately realisable and, thus, highly risky. Therefore, unless specific markets are found or created for coconut oil, its displacement by simulated oils could lead to even greater impoverishment for coconut farmers. And as always, the landless workers would be the most hard hit. On the other hand, bio-conversion of coconut could provide some solutions to expand both the local and foreign markets for the coconut. Moreover, small-scale bio-conversion systems can be designed to initiate the industrialisation of the coconut industry in the countryside.

Improvements in crop-plant production and processing attainable through biotechnology tend to promote certain trends in health awareness which eventually affect the world market for specific commodities produced by the Philippines. Looking into developments in the United States sugar market, one can see that the shrinking of that market has been triggered by the low-calorie, sugar-free diet which is being adopted out of growing health consciousness. From current developments, it appears that the Philippine coconut oil exports may suffer the same fate, owing to recent and developing concern about the effects of cholesterol intake. High blood cholesterol has been associated with heart disease, and it is generally believed that saturated fats of coconut oil are cholesterol intermediates and, therefore, raise blood cholesterol. Health concerns such as these have been reinforced by the ability of the technology sector to develop alternative products that cater to such worries and are made easily available. For example, the ready availability of artificial sweeteners reinforced the adoption of low-calorie, sugar-free foods as part of a regular diet, thereby significantly reducing sugar demand.

Biotechnology gives a tremendous boost to the ability of the technology sector of industrialised countries to rapidly fill in needs created by health fads. A growing health concern in the United States and other industrialised countries is the preference for organically grown pesticide-free food. The exports of fruits from the Philippines – banana, pineapple and mango – are the second biggest users of pesticides imported into the Philippines (Table 9.1). Given the rapid pace with which disease and pest-resistant crops suitable to new environments could be developed through tissue culture and genetic engineering, there is a possibility that even the country's world fruit market could shrink or remain stagnant rather than expand. Apparently, it would not be sufficient for the Philippines to simply diversify the export crops as recommended by the Department of

TABLE 9.1 *1984 pesticide use by crops*

Crop	Total (pesos)	pesticide[1] %	Insecticide %	Fungicide %	Herbicide %
Rice	300000	36.6	67.8	0.6	31.3
Banana	205000	25.5	0.7	62.6	9.3
Vegetables	114980	14.5	–	–	–
Mango	55880	6.9	85.9	11.7	–
Corn	36062	4.5	91.0	–	8.8
Tobacco	16620	2.1	99.2	0.1	–
Cotton	7013	0.9	95.8	2.2	–
Sugarcane	4650	0.5	14.6	5.7	74.6
Pineapple	2250	0.3	8.4	81.0	10.6
Coffee/Cacao	2000	0.27	74.5	4.5	20.5

1. Other pesticide items in the total but not itemised in the table include fishponds, fruits, snail chemicals, etc.

SOURCE: Agricultural Pesticide Institute of the Philippines, 1984.

Agriculture but also to consider trends in consumer health fads that could affect the export market. On the other hand, the benefits of using either biological control or developing pest – and disease-resistant export crops may not only be derived from the savings in foreign exchange for the purchase of pesticides, and the cleaner environment that results due to reduced use of pesticides, but it may also ensure a steady, expanding market abroad.

The new production systems of producing animal vaccines and hormones through fermentation by genetically-engineered cell systems could be beneficial and assist in attaining higher farm incomes. Since almost all veterinary products are imported into the country, the benefits of these new technologies in the form of cheaper and safer vaccines would affect producers as well as consumers. Lately, a multinational enterprise has introduced to the Philippines a human vaccine produced by a genetically-engineered micro-organism and priced less than half of its traditionally produced counterpart, thereby increasing its market towards a lower-income group. Similarly, cheaper and safer animal vaccines could become affordable by a greater number of livestock raisers, increasing production and supply and lowering prices of livestock products. On the other hand, the adoption of animal hormones to improve productivity is complicated by the need for technical assistance and efficient management abilities by the farmer-user. Given the nature of earlier experiences in the Philippines, it is more than likely that only farmers near agricultural research and extension centres will benefit from this technology.

BIOTECHNOLOGY ACTIVITIES IN THE PHILIPPINES: IS THERE
A PRO-POOR FOCUS?

The above possible adverse impacts of biotechnology developed in the
industrialised countries could be reduced if Third World countries develop
an independent capacity in key areas of advanced biotechnologies and
prevent the adoption of legislation permitting their patenting. The creation
of centres such as the International Centre for Genetic Engineering and
Biotechnology, as independent sources of knowledge for like-minded
countries, has also been seen as a countermeasure to the privatisation
of biotechnology. In addition, biotechnology has been identified as a
major area for support by quasi-public institutions and major international aid agencies. So, how does the Philippines fare in this respect?
The following section presents the trends in biotechnology research and
development in the Philippines and discusses the potential impact of
beneficial biotechnologies being locally developed.

Biotechnology capability

There is an awareness of the importance of biotechnology in the
Philippines. Formally, the Government extended support for a biotechnology research programme in December 1979, with the establishment
of the National Institute of Biotechnology and Applied Microbiology
(BIOTECH) within the University of the Philippines. The mandate of
BIOTECH is to harness micro-organisms to transform renewable raw
materials into food, fuel and fertiliser (Padolina, 1985). Prior to BIOTECH

TABLE 9.2 *Expenditure for biotechnology research*
with agricultural applications (in million pesos)

	1982	1984	1985	1986
Total expenditure for agricultural research	76.299	104.338	120.552	108.981
At constant 1978 prices	44.052	36.431	34.189	30.476
Expenditure for biotech projects	10.210	8.343	29.324	12.496
At constant 1978 prices	5.895	2.913	8.317	3.494
As percentage of total expenditure for agricultural research	13.4	8.0	24.3	11.5

SOURCE: Pooled from data obtained from BIOTECH, UPLB Office of the
Director and from PCARRD.

TABLE 9.3 *Budgetary expenditure for research at the National Institute
of Biotechnology and Applied Microbiology (in million pesos)*

	1980	1981	1982	1983	1984	1985	1986	1987
Total	8.021	5.811	10.210	6.006	8.343	29.324	12.496	12.783
Grants	–	–	2.627	–	3.27	24.902	7.119	6.581
Core	8.021	5.811	7.583	6.006	5.073	4.422	5.377	6.202
Personnel	1.287	3.178	3.257	3.227	2.243	3.556	4.620	5.187
Maintenance of operations	4.118	2.633	2.166	2.180	1.731	0.866	0.757	1.015
Equipment	2.616	0	2.160	0.599	0	0	0	0

SOURCE: BIOTECH, UPLB, Office of the Director.

there were small biotechnology projects conducted by agencies within the
Agricultural Research Network coordinated by the Philippine Council for
Agriculture and Resources Research and Development, but these have
not been identified separately as such but rather as components of
commodity-oriented programmes in agriculture. Consequently, the support
for biotechnology research programmes for application to rural areas has
not been consistent (Table 9.2). This seems to reflect the general lack of
government commitment to research and consequently the dependence of
the research sector on external grants. Without the external grants, there has
actually been a continuing reduction in support for biotechnology research
(Table 9.3). More disturbing is the consistent decrease in allocations for
maintenance and operating expenses.

Nevertheless, the scientific community has maintained a rural sector-
oriented programme. BIOTECH has adopted four inter-disciplinary
research programmes, namely: bio-fuels from agricultural crops and
residues, nitrogen fixation and enhancement of soil nutrient availability,
establishment and improvement of traditional food fermentation processes,
and special projects on veterinary antibiotics, plant diagnostic and plant
cell cultures for secondary metabolite production. For the years 1980–84,
BIOTECH spent 26 per cent of its direct costs of research for the bio-fuels
programme, 32 per cent on nitrogen fixation programme, 19 per cent on
the food programme and 23 per cent for its special projects. On the other
hand, the Agricultural Research Network has been consistently active in
tissue-cultured seeds and bio-fertilisers (Table 9.4).

The support for biotechnology is expected to increase as two government
agencies within the Department of Science and Technology come into
operation – the Philippine Council of Agriculture and Resources Research
and Development (PCARRD) and the Advanced Science and Technology

Institute (ASTI), both having mandates to promote biotechnology research and development. In May 1988, the PCARRD included agricultural biotechnology as one of its four priority sectoral concerns, and identified genetically-engineered planting stocks as a priority commodity concern for research. The long-term approach of producing restriction fragment length polymorphisms (rflp) maps for major crop plants is considered a top priority.

There is private-sector involvement in biotechnology. One seed company in the Philippines, a subsidiary of a multinational enterprise, has started research on corn tissue culture. A private company teamed up with the University of the Philippines at Los Baños for the testing and commercialisation of *Paecilomyces lilacinus*, a fungus against root nematodes attacking banana, potato and other crops. Negotiations are currently under way for a joint sugarcane biotechnology programme between scientists of the University of the Philippines, sugar millers, sugar planters and the Sugar Regulatory Administration.

The issue of bio-safety has been raised and pending before the Philippine Senate is a bill creating a national Biosafety Board for the regulation of biotechnology projects. Parallel to the Senate move is the creation of an interagency committee on bio-safety created by the heads of the Department of Agriculture, the University of the Philippines at Los Baños

TABLE 9.4   *Relative expenditures in various research areas in biotechnology within the Agricultural Research Network (in % of total expenditure for agricultural biotechnology)*

Research area	1981	1982	1984	1985	1986	1987
Tissue cultured seeds	23.3	21.9	21.9	20.2	34.8	41.0
Bio-fertilisers	22.6	11.0	21.6	28.8	44.8	33.5
Microbial pesticides	5.1	19.9	6.3	6.1	5.4	–
SCP	32.0	30.7	31.6	19.4	–	2.9
Bio-fuels (alcohol and biogas)	6.9	12.8	0.2	13.2	3.7	3.4
Bio-conversion for food beverages (including mushrooms)	–	–	5.1	9.8	4.7	10.9
Plant growth regulators	6.9	–	0.9	–	–	1.0
Diagnostics (plant)	–	–	12.4	2.4	6.6	8.3
Animal reproduction (superovulation)	3.1	3.6	–	–	–	–

SOURCE: Calculated from data supplied by the PCARRD, Office of the Director.

and the International Rice Research Institute (IRRI). Guidelines on the conduct of biotechnology projects (research and non-research) have been formulated by this committee.

Although the issue on bio-safety has been initially raised from circumstances arising from laxity of quarantine procedures and unfounded fears of biotech projects, the concern of others could be real. There may be a need for such a law considering the experience of the Philippines as the dumping ground of unwanted technologies and products from developed countries. The possible use of the Philippines as experimental ground for biotech products and processes also cannot be discounted. Due to the lack of experience in this field, there is no knowledge of the risks involved in biotech products. Understandably, the PCARRD has included bio-safety as a researchable area.

The research techniques currently employed in the development of various biotechnologies have not quite reached the level of sophistication one automatically associates with genetic manipulations. Most of the work being done with N-fixation is basic microbiology and field testing, biogas concerns techno-feasibility, alcohol research involves process development, bio-pesticides constitute basic microbiology and efficacy testing and tissue-culture is used for micropropagation.

Of these biotechnologies, tissue cultured seeds appear to be moving forward technically, in that 13 out of 60 studies reported are now concerned with cell and protoplast culture and *in vitro* selection. Of about 100 research projects examined, only two projects, one on coconut disease and another on cellulose degradation, are working at the DNA level, and four projects use cell fusion as a technique for improving cellulose degradation, for increasing alcohol yields, for improving animal vaccine production and/or producing monoclonal antibodies for plant viral diagnosis. The focus of these projects reflects both the limited number of personnel trained in the new technology and the lack of funds to support new projects.

Women biotechnologists

Interestingly, work on all of these advanced projects is carried out primarily by women. And, on the whole, based on membership of professional societies, the field is dominated by women. The Philippine Society for Microbiology consists of 80 per cent female and 20 per cent male members, the Cell/Molecular Biology and Biotechnology Society is 74 per cent female and 26 per cent male, and the Philippine Association for Plant Tissue Culture is 85 per cent female and 15 per cent male. The

predominance of women in the field could be related to character traits
that appear to be gender associated. Cell and molecular biology research
requires patience and perseverance, traits often associated with Filipino
women rather than men. Primarily, the reason could be economic. That
is, scientific research in the biotechnology field in the Philippines does
not pay well, hence the Filipino male who is expected to be the family
provider shies away from a scientific career in this field.

LOCALLY-DEVELOPED BIOTECHNOLOGIES

Local research resources have been expended mainly on the development
of four biotechnologies, namely, bio-fertilisers, bio-fuels, bio-pesticides
and tissue-cultured seeds. Each of these is examined more closely below
to assess possible impact particularly on the small farmers.

TABLE 9.5   *Actual and projected prices for fertilisers*
(FOB prices at constant 1986 dollars)

		$/ton		
	Urea (North West Europe)	Diammonium phosphates (Florida)	Triple superphosphate (Florida)	Potash (Vancouver)
Actual:				
1972–76 Crisis				
1972	141	216	160	79
1973	212	266	223	95
1974	651	686	626	125
1975	373	458	381	153
1976	201	215	163	99
1986	107	160	120	73
1987	112	150	126	68
Forecasts:				
Annual				
1988	130	185	135	70
1989	140	200	150	70
1990	180	205	155	70
1991	220	225	170	70
1992	280	265	210	75
1993	260	325	260	85
1994	230	265	210	90
Trend	220	250	195	125

SOURCE: Fertiliser and Pesticide Authority, March, 1988.

TABLE 9.6   *Nutrient removal by crops*

Crop	Yield level (Kg/ha)	Nutrient N	removal P	Kg/Ha K
Rice	4000 (grain)	90	35	110
Corn	4500 (grain)	110	45	155
Sugarcane	90 000 (canes)	120	75	250
Coconut	100 (trees)	53	50	93

N = Nitrogen; P = Phosphate; K = Potassium.
SOURCE: Fertiliser and Pesticide Authority, 1988.

Bio-fertilisers and labour intensity

There is a strong particular interest in harnessing biological nitrogen fix-ation in producing bio-fertilisers in the Philippines. One reason is that the Philippines imports all of its chemical N-fertiliser and additional foreign exchange expenditure is required to fully supply the needs for expanding agricultural production and to meet the additional costs resulting from the projected increase in fertiliser prices (Table 9.5). Moreover, increased use of fertiliser is needed because crop production depletes the soil of nutrients (Table 9.6). More importantly, bio-fertilisers are perceived to be cheaper and within the financial reach of most Filipino farmers who cannot maintain their level of fertiliser requirements due to financial constraints.

Systems recommended and studied as bio-fertilisers include the *Azolla-Anabaena*, crop legume-*Rhizobia*, *Sesbania-Rhizobia* and *Leucaena-Rhizobia* as green manures or compost components, compost, blue-green algae and nonsymbiotic nitrogen-fixing bacteria. Recipient crops being studied are crop legumes, mainly peanut, mungbean and soybean, rice, corn and sugarcane. Rice and sugarcane production consume 78.4 per cent of the total fertilisers traded in the country. Corn is the most extensively grown crop by small farmers in the nutrient-poor soils of the uplands and yet, fertiliser application is not a common practice.

A rice, corn or sugarcane farmer has several bio-fertilisers to choose from. An upland rice or corn farmer has a choice among a legume (*Leucaena*) green manure, compost and a nonsymbiotic N-fixing bacteria. A lowland rice farmer has the *Azolla-Anabaena* system, *Leucaena* or *Sesbania* green manure, a free-living N-fixing and blue-green algae to choose from. All these bio-fertilisers are simple to produce and do not require sophisticated skills and equipment. However, they do require different amounts of resources. Table 9.7 compares the resource require-ments of producing and using four bio-fertilisers equivalent to 30kgs N;

TABLE 9.7 *Main inputs needed to produce biofertilisers equivalent to 30kg N/ha current price, 1988 (market price @ $0.09/kg urea)*

| | COMPOST | | | | AZOLLA | | | |
| | Quantity | Unit cost | | Total cost | Quantity | Unit cost | | Total cost |
		Actual	Breakeven			Actual	Breakeven	
Labour	64 mhrs	$0.28/mhr	$0.09/mhr	$17.92	40/mhrs	$0.28/mhr	$0.15/mhr	$11.20
Materials Seeds/Inoculum required	32kgs	$0.48/kg	–	$15.36	–	–	–	–
Fertiliser	0	–	–	–	12kgs Phosphate	$0.16/kg	–	$1.92
Pesticide	0	–	–	–	.015–2kgs	$37.00/kg	–	$0.56–$74.00
Opportunity cost	1400kgs	$0.01/kg Leucaena leaves	–	$14.00	–	–	–	–
	1400kgs	$0.02/kg Chicken manure	–	$28.00	–	–	–	–
Area taken out of production to produce biofertiliser	0	–	–	–	–	0.004–0.1 ha (for inoculum production)	–	–
Time required to produce effective biofertiliser at the farmer's level	28–30 days	–	–	–	–	20 days	–	–

TABLE 9.7 *continued*

	LEUCAENA				NON-SYMBIOTIC N-FIXING BACTERIA		
	Quantity	Unit cost		Total cost	Quantity	Unit cost	Total cost
		Actual	Breakeven				
Labour	120 mhrs	$0.28/mhr	$0.05/mhr	$33.60	2/mhrs	$0.28/mhr	$0.56
Materials							
Seeds/Inoculum required	0.875kgs	$0.95/kg	–	$0.83	1 packet	$0.48	$0.48
Fertiliser	6.25kgs	$0.16kg	–	$1.00	–	–	–
Pesticide	–	–	–	–	–	–	–
Opportunity cost	1500kgs dried Leucaena	$0.10/kg	–	$150.00	–	–	–
Area taken out of production to produce biofertiliser	0.5ha	–	–	–	0	–	–
Time required to produce effective biofertiliser at the farmer's level	90 days	–	–	–	–	–	–

Azolla-Anabaena as in situ grown green manure, *Leucaena* as cut and carry green manure, and nonsymbiotic N-fixing bacteria. The N-fixing bacteria requires the least of the farmer's resources provided that the inoculum is produced and made available to the farmer at the current cost of *Rhizobium* inoculants. A good measure of government investment is required in setting up a network of fermentation plants to provide this service. This also means that the system should improve upon that presently used for Rhizobium inoculum production and distribution, which is not even used in more than 10 per cent of the areas planted to legumes. It should be noted that this technology addresses the most disadvantaged farmer groups, the upland rice and corn farmers. The technology also addresses sugarcane farming, the most intensive user per unit area of N-fertiliser based on PHILSUCOM recommendation of 200–100–300 NPK per hectare.

The bio-fertilisers are labour-intensive, especially *leucaena* cut and carry, green manuring and composting. Although it is often believed that labour-intensive technologies are economically desirable in the rural areas to absorb the increases in rural labour supply, those technologies that support production of price-controlled commodities such as rice would per-petuate rural poverty and widen the gap between the landless poor and the rest of the rural sector. This would be true in areas where labour supply is high and 'gama' is practised. 'Gama' is the practice of non-payment in cash of labour supplied during the growing phase of crop production. Instead, all the hired labourer gets is a free meal and an assurance that he/she will be part of the harvesting and threshing team (Aragon *et al.*, 1985). 'Gama' therefore is exploitative of landless workers, since essentially the labour rendered before harvest is either free or loaned out to a resource-richer member of the sector. Labour-intensive technologies tend to reinforce 'gama' which is practised in some regions of the country. On the other hand, if sufficient family labour exists and no alternative non-farm income generating activity is available, the use of compost, *Azolla* and *Leucaena* would make economic sense only when the cost of the compost inoculum is lowered significantly below the price of the substituted N-fertiliser. The *Azolla* requires less or no pesticide, and *Leucaena* leaves cannot be sold as feed component and its production does not require pesticides. On the other hand, research is currently under way on a system of green manuring using *Sesbania* with much reduced labour requirements.

Tissue culture directed to small farmers

No conscious effort is locally made in directing biotechnology develop-ment to different types of farmers. Virus-free root crop stocks produced by tissue-culture are available at almost no cost. But embryo-rescued mutant coconut (macapuno) planting stock is too expensive ($23.80/plant) for all but plant collectors to buy.

The International Rice Research Institute has done intensive work on rice biotechnology supported by the Rockefeller Foundation and is directed towards rice farmers missed out by the Green Revolution. However, only a few of the products of this technology development are applicable to Philippine conditions. This is also true of the Tissue Culture for Crops Program of the Colorado State University which is supported by the United States Agency for International Development. The latter programme, however, does support upland rice and corn improvement through a tissue culture project, which is being carried out at the Visayas College of Agriculture in Central Philippines.

What this indicates is that international research programmes should not be looked upon as sufficient to generate location-specific biotechnologies, and that local capabilities need to be developed to fully utilise technical assistance and training provided by these international programmes.

Bio-pesticides in the Philippines

Although bio-pesticides are helpful to farmers and to the public in general, since they are non-pollutant as compared with chemical pesticides and therefore less environmentally damaging, bio-pesticides like chemical pesticides may differentiate between poor and rich farmers. Since the trend is to develop a capital-intensive production technology, the production, distribution and sale will be in the hands of the private sector. The University of the Philippines has recently awarded the production of a fungal nematicide for banana, potatoes and other crops to a private company. Although this nematicide is priced at least 30 per cent lower on an area need basis than its chemical counterpart, this pricing scheme is evidently directed not towards market expansion but rather more towards product promotion. The price could have been much lower, and affordable to more farmers, had the pricing been based on production cost. Thus, unless more production firms enter to ensure competition, improve efficiency of production and bring down prices, the real benefits of such locally-developed technology will not spread to a greater number of farmer-users.

Bio-fuels: Potentials

Bio-fuels in the present Philippine context are biogas, the flammable, odourless gas produced by anaerobic micro-organisms actively growing on agricultural or industrial wastes, and fuel alcohol, anhydrous ethanol derived from the fermentation by yeast of sugar-rich materials. The technology development for biogas has been directed towards the richer farmers because of the capital outlay needed to make the system operational. The technology is subject to economies of scale and although efforts have been made to lower the capital cost, the smallest system is still too expensive at $286–332 per unit, not including the investment cost of rearing pigs to supply the raw materials. Presently, large industrial plants and livestock farms are more interested in biogas systems for their waste treatment value, and with their fuel value as an added bonus. The fuel value, however, is extremely attractive with the investment for a system paying off about two years after the start of operation and continuously saving on fuel/electricity costs thereafter. Maya farms, for instance, a 6000 head livestock farm and food-processing complex, rely completely on biogas for their fuel needs, saving the company an equivalent value of about $38 000 monthly on electricity costs.

RESEARCH STRATEGY AND PLANNING

From the foregoing discussion, it is apparent that the current trend in biotechnology development abroad could adversely affect Philippine agricultural development and that the national efforts of harnessing new knowledge for agricultural technology development lack adequate support, direction and focus.

Sugar may face a stiffer competition from new sweeteners such as thaumatin or from a cheaper HFS and, similarly, coconut from simulated oils and fats and fruits, from pesticide-free fruits. The poor corn farmers may even lose their chance to sell in the local feed market with stiffer competition from imported biotechnology-based corn.

Although biotechnology R&D activities have been launched in the Philippines and such activities are directed to existing problems, very few are directed towards the more pressing problems of survival of the farm population and maintenance of the comparative advantage in the Philippines' major crops. Furthermore, the national capability to use the revolutionary techniques of cell and molecular biology is very limited, and such limited capability is spread out thinly to various unrelated problems.

Biotechnology is location-specific and requires local capability for development. Thus, it is essential that the Philippine public perception towards its own scientific sector changes. And at the same time, this scientific sector must also realise that biotechnology is a strategic technology, central to the struggle to retain or regain international competitiveness, and certain to have a major economic impact.

With the rapid pace of biotechnological developments, Philippine agriculture needs to adopt a growth plan that must take into account these developments and adjust its strategies accordingly. Undoubtedly, biotechnology must be part of these strategies as it is a major tool that can provide that quantum leap from scarcity and vulnerability to abundance and stability.

10 Can We Avert an Oil Crisis?

Guido Ruivenkamp

INTRODUCTION

Biotechnology reinforces the processes of interchangeability for all the different basic nutrients such as carbohydrates, proteins and oils. This restructuring process leads to an increase in the interchangeability of product ranges and manufacturers, and as a consequence, to a decline in prices. Developing countries cultivating and exporting crops will in the first place be confronted with the consequences of the introduction of biotechnology elsewhere. The necessity to restructure and adapt will be felt most urgently, especially in these countries.

In this chapter the impact of increased flexibility in the supply of raw materials to the food-processing industry will be analysed for the production of oils and fats globally. The first section deals with the present-day interchangeability of oils and fats, which is limited to those sources belonging to one of the four major groups of oils with an almost identical composition of fatty acids.

In the second section three major developments concerning biotechnology applications to the production of vegetable oils are described. Through the applications of these biotechniques the farmers in several countries will be affected because the boundaries between the four groups of oils gradually disappear. The result of plant breeding techniques, enzymatic applications and the microbial production of fatty acids is a changeover to a new production system.

In the third section, three general aspects of the restructuring process are dealt with: a new form of labour organisation, enzyme manufacturers as 'political agents', and regional self-sufficient food systems as interchangeable production units.

The final section analyses the consequence of the international restructuring process for the oil-producing regions. Special attention is paid to the reorganisation of large – and small-scale farming systems of several vegetable oil sources within this new world system of food production. The restructuring of palm oil production is the context in which this new position of farming systems is dealt with. The changeover to a new food system, in which large-scale and small-scale farming systems are increasingly considered as interchangeable as production units, will also modify the political meaning of the initiatives to extend the diffusion of biotechnology to small-scale farming systems.

Interchangeability of oils and fats

Vegetable oils and animal fats are important raw materials which are used in food and non-food products. In the food-processing industry already a broad spectrum of different oil sources (80 per cent vegetable oil sources) are used, ranging from soyabean, palm, sunflower seed, rapeseed, arachide to coconut, palm kernel and cotton seed. From the more than 11 vegetable oil crops, soya, palm, sunflower, rapeseed, coconut and palm kernel oil are competitors on the world market. Out of a global production of 63 145 million tons, approximately one third (25 538 million) was exported with a leading position for two crops in these exports, soya (46 per cent) and palm (22 per cent).[1] The dominant position of soya oil on the world market is determined partly by its specific fatty acid composition. The relatively high degree of linoleic and linolenic acids makes soya more attractive for uses in food items by the oil-importing countries than coconut oil. The second reason for its dominant position is that soya oil is a residue from soya protein production. Therefore, the price of soya oil can be kept to a minimum and only those oil crops with high yields per hectare, such as palm oil, are able to compete with it on the world market. The intensive competition between soya oil and palm oil has also put under pressure the production of other vegetable sources, like the arachide oil production in Senegal and coconut oil production in the Philippines.

Next to the general uses of the different oil crops (for which the price is the decisive factor), there are also specific markets in which one type of oil is more convenient for use by industry than another. The different fatty acid composition results in a comparative advantage of certain oil sources for specific uses. For example, coconut oil contains a relatively large quantity of lauric acid, palm oil oleic acid and soya oil linolenic acid. In the same way, arachide oil is relatively more attractive for some specific food purposes than palm oil because of its higher linolenic acid content. However, higher yields of palm oil could reduce this natural advantage of arachide oil.

TABLE 10.1 *Interchangeability of oils and fats divided in four groups with an almost identical fatty-acids composition*[1]

		Characteristic acid group and other acids in %						
		Group 1	Group 2	Group 3	Group 4			
Region/country	Oil Source	Triaglyc. Linolenic C18:3	Diglyc. Linol C18:2	Monoglyc. Oleic C18:1	Single saturated Stearic C18:0	Palmitic C16:0	Myristic C14:0	Lauric C12:0
EC	Rapeseed	6–14 (1)						
Brazil	Soja	4–11 (2)	44–62 (1)	19–30 (4)		7–14 (4)		
United States	Corn	2 (3)	34–62 (2)	19–50 (3)		8–19 (3)		
EC	Sunflower		20–75 (3)	14–65 (5)				
Senegal	Arachide		13–45 (4)	20–75 (2)		6–16 (6)		
Mali	Palmoil	1.5 (4)		27–52 (1)	2–8 (2)	32–59 (1)		
Mali	Palm kernel			10–23 (6)		7–11 (5)	14–20 (2)	42–55 (1)
EC	Butter				28 (1)	13 (2)	24 (1)	10 (3)
Philippines	Coconut						13–23 (3)	41–56 (2)

1. Some acids are not dealt with like butteracid (C4:O), caprinic acid (C10:0) and erucinic acid (C22:1), which is not adapted for food-uses and especially present in rapeseed.

The vegetable oil sources can be divided into four groups – the triaglycerols, the diglycerols, the monoglycerols and the saturated oils and fats (see Table 10.1). The interchangeability of vegetable oil sources is actually still limited to those sources with an almost identical composition of fatty acids. For the triaglycerols, the producers of soya oil and rapeseed oil are interchangeable suppliers of linolenic acids. Moreover, the surpluses of corn oil and palm oil might have a price-depressing influence on the supplies of these acids.

For the oils with a high degree of linoleic acid (the diglycerols), the food-processing industries can choose from a large number of different sources like soya, corn, sunflower, and arachide.

The interchangeability of the producers is extremely high for the production of oleic acid (the monoglycerols). Several sources might be used. Moreover, the high palm oil yields could cause the price of oleic acid production to drop. In the fourth group of saturated oils, there is competition between vegetable oil sources and animal fats, although different sources do have a comparative advantage, like butter for the stearic acid and palm kernel oil for the extraction of palmitic and myristic acids. There is strong competition between coconut oil and palmkernel oil for the production of lauric acid. Interchangeability of the vegetable oil producers within the boundaries of these four groups is illustrated in Table 10.1.

The position of farmers in several countries producing these vegetable oil sources might be affected by biotechnological developments. Biotechnology may result in higher yields of several crops. Moreover, the fatty acid composition of several vegetable oil sources might be modified.

BIOTECHNOLOGY APPLICATIONS IN OILS

Biotechnology can increase the flexibility in the supply of oil sources to the food-processing industries through three developments: (1) The development of better extraction techniques and the enzymatic modification of oils and fats, such as fractioning, modifying and transferring fatty acids; (2) the development of plant biotechnology; and (3) the microbial production of oils and fatty acids. Through the applications of these techniques the interchangeability of producers will no longer be limited to sources with an almost identical fatty acid composition given that the differences between the four groups will gradually disappear. We will discuss in which way and by whom this specific introduction of biotechnology in the global production of vegetable oils is going to be organised.

Extraction techniques and enzymatic modifications of oils and fats

In addition to the possibility of developing new enzymatic methods of extracting oils from plant seeds, the impact of enzyme technology depends on the possibility of modifying the fatty acid composition of vegetable oil sources. Enzymatic modification of oils and fats can be achieved in three ways:

(a) The development of new lipases which are able to fraction oils. The enzymes might be able to displace some specific fatty acids or to break up the triaglycerols into di- or monoglycerols. By splitting up certain fatty acids the utilisation of high-yielding oil sources might increase. For instance, the smell or taste of oils can be modified by isolating certain acids but through deodorisation coconut oil can be adapted to the taste of the Western consumer and be used as baking oil.

(b) Esterification of oils is concerned with the building up of monoglycerols into di- and triaglycerols. Lipases can also be used to add certain fatty acids. By this synthesising technique monoglycerols can be built up into di- or triaglycerols. In this way cheap oil sources with the addition of certain desired acids can become competitors for the more exclusive oil sources.

(c) The mutual transfer of fatty acids constitutes the third method of breaking down the boundaries between different groups of oils through a mutual exchange of fatty acids. Through the transfer of fatty acids, cheap sources of oils (corn, palm) can become the raw material for the more expensive oil sources. Specific components of oils and fats can also be replaced by a selective transfer of fatty acids. This technique could be applied in cacao butter production.

The food industry is already using a large number of substitutes for cacao butter. These are usually composed of a mixture of cheap palm oil fractions together with expensive fractions of exotic oils. Through the inter-esterification of palm oil and stearic acid, biotechnology makes it possible to obtain a cacao butter equivalent whereby the expensive oil fractions are no longer needed. In addition to the possibility of replacing cacao butter by palm oil with the addition of stearic acid, it also becomes possible to produce cacao butter substitutes in which the expensive oil fractions from countries like Burkina Faso, Mali and Benin are replaced by EC butter surpluses. Unilever has financed this research on selective fatty-acid exchange, especially the inter-estification of palm oil into a cacao butter product.

In short, developments in enzyme technology will increase the possibilities of breaking up the triaglycerols into di- and monoglycerols, and of building up the monoglycerols into di- and triaglycerols as well as the possibility of transferring fatty acids between different groups of oils and fats. As a consequence the boundaries between different oil sources will disappear. The position of farmers in the vegetable oil-producing countries will be affected because they become interchangeable, with an increased number of producers of other (vegetable and microbial) oil sources.

The reorganisation of the world production of oils, with increasing flexibility in the substitution of one oil for another, also tends towards a situation in which the production of oils could be privatised. For example, the Japanese company Suntory Limited applied for a patent (20–7–84, J) on the production of lipases from a breeded microorganism (Staphylococcus capitis). Exclusive property rights are claimed on the use of this enzyme to convert undesired fatty acids (European Patent Office, EPO, Nr.0.187.869 A1 WO 86/00925). The conversion of low quality oils into higher-quality oils could be realised by this enzyme. In a second patent (20–7–1984), Suntory Limited has also claimed the method of using this new microorganism for the microbial production of fatty acids which might be used in non-food sectors (chemicals, medicines, perfumes, cosmetics, (EPO Nr.0.188.628 A1 WO 86/00932). Another Japanese company, Sapporo Breweries Limited, has applied for a patent (5–6–1986) for a new enzyme (produced by *Pseudomonas fragi* 22.39B) to split up fatty acids from a broader spectrum of oils and fats than the actual number of existing enzymes (0.204.284 A2). In addition, Bellex Corp. (J) has patented the development and use of new lipases which can function on a broader substrate composition. The scope the enzyme's use can be extended, with the result that a certain fatty acid composition can be produced by differentiated conversion routes using the same patented enzyme (0.149.520 A2). Various production systems could become dependent on the supply of that enzyme.

Another possibility of increasing the influence of the lipases producing companies is the development of lipases which are able to function under specific temperature conditions. An example is the Japanese company Daikin Kogyo Co. Ltd. which has applied (25–2–1983) for a patent on the development and use of heat-resistant lipase (from Rhizopus chinesis IFO 4745).

Plant biotechnology

Besides fractioning, synthesising or transferring fatty acids, the inter-changeability of different vegetable oil sources might be increased by the development of plant biotechnology. Through the application of several new techniques, such as somoclonal and gametoclonal propagation, protoplast fusion, the time needed for developing new varieties of annual crops (soya, rapeseed, sunflower, cotton) will be reduced from 7–8 years to 3–4 years, and for the perennial plants (like palm oil from 30–40 years to 7–8 years). New high-yielding varieties, resistent against herbicides or plant diseases, as well as new varieties with a modified fatty acid composition might be realised sooner. Table 10.2 shows different examples of plant breeding programmes.

Within the countries of the European Community (EC), initiatives have been taken to change the fatty acid composition in rapeseed. An attempt is being made to lower the erucinic acid and to raise the degree of linolenic acid. Through these modifications rapeseed might be used more frequently for food-processing purposes within the EC. Reductions of soya oil imports into the EC could be the consequence of these plant breeding programmes.

Several companies and institutes, using clonal propagation techniques, are developing new palm oil varieties with higher yields (see the final section of this chapter). An increased palm oil production will also lead to higher palm kernel production and, in turn, will influence the position of the coconut oil producers in the Philippines, especially in the small-scale farming systems.

Palm kernel and coconut oil are interchangeable sources for the saturated oils (see Table 10.1). Soya research in the United States is concentrated on lowering the linolenic acid as well as increasing the oleic and stearic acid content. By these modifications soya oil will become more adapted for technical applications. The relative advantage of palm oil – due to its relative higher degree of oleic acid – will be diminished. The competition between soya oil and palm oil will be intensified and producers of many vegetable oil sources might be put under pressure (see the third section).

Deploying microbes

Not only plants and animals but also microorganisms are able to produce oils and fatty acids which are comparable to the vegetable oils in their composition. The main condition which has to be satisfied is that the organism grows and becomes fatter instead of multiplying. This can

TABLE 10.2 *Plant breeding techniques and increased flexibility in the oils sector*

Crop	Shorter breeding time	Higher yields	Research targets		Modified Institute composition	Research
			Area-extension	Resistant fatty-acid		
Rapeseed	From 7–8 to 3–4 years	+	EC Winter persistent	Disease-R Herbi-R (triazine	Higher linolenic (C18:3) Lower erucinic (C22:1) Higher erucinic (C22:1) for technical purposes	EC and Biotechnica Intern Inc. (Canada)
Soya	From 7–8 to 3–4 years	+	Colder regions Inter-seeding with wheat	Disease-R Herb.-R	Lower linolenic (C18:3) Higher oleic (C18:1) Higher stearic (C18:0)	Monsanto (US)
Palm oil	From 30–40 to 7–8 years	+	Indonesia Brazil	Disease-R	Higher oleic (C18:1) Higher linoleic (C18:2)	Unilever
Sunflower	From 7–8 to 3–4 years	+	EC		Higher oleic (C18:1)	Higher linolenic (C18:3) EC & Sungene Techn.
Coleseed	From 7–8 to 3–4 years	+			Higher saturated acids (C12) for detergents	
Coconut oil	From 30–40 to 7–8 years	+	Philipines (Japan)			Unilever and Kao Corp.

SOURCE: The table is based on data published by Tanis 1987, 37. An institution which is occupied with the above-mentioned research is the American Soybean Association, Arco.

be realised by a certain medium composition in the fermentor which brings the microorganism to convert the excess carbohydrate substrate into lipid without further cell proliferation. Plant cells, algae, bacteria, yeasts and moulds are therefore 'fattened'. However, like plants, not all microorganisms accumulate high amounts of lipid. This property is shared by only a relatively small number of the total number of species (known as the oleaginous species). Besides, some oleaginous species like *Mycobacteria*, *Corynebacteria* and *Nocardia* are well known for their high lipid content but these lipids tend to be associated with toxic and allergic factors. Yeasts and moulds are actually the most efficient organisms for producing lipids. However, they accumulate no more than 20 per cent of their biomass as lipid. Nevertheless, the fatty acids of the oleaginous yeasts and moulds show their similarity to many plant seed oils. The predominant fatty acids are oleic acid, palmitic acid, linoleic acid, stearic acid and palmitoleic acid. However, some genera will produce in a different order of abundance and some organisms may produce relatively large proportions of linolenic acid (18:3). Selecting specific organisms might become very important for the production of microbial oil. For instance, the moulds, *Entomorphtora coronata* and *Entomorphotora obscura*, produce an abundance of lauric acids that are normally obtained from coconut and palm kernel oil sources.

Microbial production will also become more important if microorganisms are able to accumulate over 40 per cent of their biomass as lipids are developed, especially if cheap raw materials such as glucose can be used or if microorganisms are able to accumulate lipids on different substrate compositions. An example is the yeast *Candida curvata*, which is able to ferment a wide range of substrates. Besides its ability to ferment the lactose in whey, the organism can utilise agricultural and food-processing wastes as substrates for fat fermentation.

In certain regions the economic competitiveness of oil fermentations might improve by using carbohydrates that have a low or even negative value, as those found in the waste products of agricultural production and food processing. In this way microbial production of oils might become a residue product of an environmental process directed to clean up water.

Another important stimulus for microbial oil production is the selection or development of a microorganism in which the lipid accumulation is high. Therefore, research is also directed towards the manipulation of micro-organisms to increase the accumulation of lipid. Recent work in laboratories has indicated that oleaginicity may be possible to define biochemically in terms of the possession of the enzyme ATP-citrate lyase which generates efficient lipid accumulation rates (Ratledge, 1984).

Research has been carried out to introduce the genetic determinants for ATP-citrate lyase in organisms which are easily manipulated.

Besides, some bacteria are known for their high lipid content. For example, the bacteria Arthrobacter AK 19 can contain up to 80 per cent of its biomass as lipid. Moreover, this lipid is composed predominantly of triacylglycerols (90 per cent). The only apparent drawback with this organism is its rather slow growth rate but it is not inconceivable that this could be improved in the near future (Ratledge, 1984). The interest in Arthrobacter AK 19 as a potential source of fats and oils was also stimulated because of the existence of a cell-wall lyzing enzyme (Novozym 234) which can be used for the extraction of lipid.

Even microalgae might achieve a high lipid content from 15 to 50 per cent. Moreoever, these algae-lipids contain predominantly triacylglycerols. However, these oils often have a fishy taste which make them unsuitable as a food product. Attempts are being made to reach an acceptable food product by mixing the algea-oils with rapeseed or palm oil.

The potential of this algal production is enormous. Compared to areal yields from soybeans, microalgae could produce up to 30 times more oil. Moreover, algal crops offer many possibilities for further enhancing oil production because cell generation times are short and thus physiological manipulation can be substantial. Also, significant variations in the content and quality of oil among different species suggest several possibilities for genetic optimisation of oil-producing species.

It is probable that from several groups of organisms efficient means of producing oils can be found outside of conventional agriculture. In Table 10.3 an overview is given of some examples of the microbial production of fatty acids which are also present in the vegetable oil sources.

Yeasts, moulds and algae can produce a variety of different oils, many of which could be used as substitutes or supplements for existing commercial plant seed oils. Yeast and moulds contain similarities in the composition of oils to the vegetable oil sources. They contain a large amount of oleic acid (18:1), linoleic acid (18:2) and palmitic acid (16:0), the same acids which are also present in palm, soya and coconut oil. Even triaglycerols such as linolenic acid (18:3) are produced by yeasts and moulds, while the di- and monoglycerols and saturated fats have been produced less.

Microbial oil production is important not only because micro-organisms contain lipids with a composition almost identical to the vegetable oil sources but also because substrates might be found almost everywhere at local level.

In addition to microbial oil production – or single cell oil production

TABLE 10.3 *Microbiological production of fatty acids*

Oil Source	Group 1 Triaglyc. Linolenic C18:3	Group 2 Diglyc Linoleic C18:2	Group 3 Monoglyc. Oleic C18:1	Group 4 Mono saturate d acids		Myristinic C14:0	Lauric C12:0
				Stearic C18:0	Palmitinic C16:0		
Algae							
Chlorophyceae	6–43	2–35	3–37	1–11	7–30	2–30	–1
Bacillariophyceae	1–3	5–16	36–62	3–15	16–36	1	–2
Yeasts							
Candidacurvata D	–	8	44	15	32	–	–2
Moulds	1	2–42	4–48	2–16	9–37	1–8	–2
Fusarium moniliforme	1	42	30	11	14	1	–2
Bacteria							
Arthrobacter AK 19	–	0.3	23.5	14.8	42.6	3.4	–3

Characteristic acid group and other acids in %

– another use of oil-accumulating organisms is the possibility of modifying oils and fats. Instead of using enzymes it might be possible to alter oils by fermentation with fungi: for example a fungus could be used which is capable of digesting and absorbing low-grade fats and oils while at the same time accumulating and depositing a valuable and desirable oil. Another example is the use of yeast as catalysator. Chloresterol might be removed from animal fats by digestion and recovery from the yeast *Candida lipolytica.*

Although microbiological processing and production cannot yet compete with vegetable oil production, microbial oils might become supplements in oil mixtures to be used for general applications. This could result in surpluses of vegetable oil sources and strong price fluctuations. For example, microbial linoleic acid might be added to palm oil, which could reduce the competitive position of soya oil on the world market. Microbial-produced lauric acid added to sunflower oil might reduce the import of palm kernel oil and coconut oil. The inter-esterification of palm oil with stearic acid might substitute cacao butter. In addition, soya oil, supplemented with microbial produced oleic acid, could put pressure on the position of other oils and increase the palm oil surpluses in certain regions.

These possibilities of combining fatty acids obtained from several microbial sources with vegetable oils will lead to a more flexible supply of oils and fats to the food-processing industry. Microbial production, enzymatic applications and plant breeding techniques could result in the restructuring of the world oil production in which the boundaries between different agricultural vegetable sources will gradually fade away.

The changeover to a new production system will alter the economic and political positions of the various economic and political actors. Three general aspects of the restructuring process will be dealt with: (a) a new form of labour organisation; (b) enzyme manufacturers as 'political agents'; and (c) regional self-sufficient food systems as interchangeable production units.

GLOBAL RESTRUCTURING OF VEGETABLE OIL INDUSTRY

The development of plant biotechnology, the use of new enzymatic extraction techniques and the microbial production of food components will increasingly modify the basic principle of food production.

The linear integration of agriculture and food processing, in which a specific agricultural product is processed into a specific end-product, no longer models the interaction between agriculture and food-processing industry. In principle, enzymatic extraction techniques make it possible to obtain food components from a broad spectrum of sources.

The overall tendency of these restructuring processes is that food products can be considered less as the result of processing one specific crop. Agricultural products will be increasingly conceived as a totality of biochemical components which might be used as general inputs for the food-processing industries. The new production system of food tends to manufacture a number of nutrients like fats, proteins and sugars, which are assembled from a broad range of agricultural products. Additives like flavouring and colouring agents are added and this packet of ingredients is put together in a recognisable form of a traditional food-stuff and sold to the consumers. This restructuring has modified the position of agricultural sectors of different countries and will lead to a new organisation of labour.

New forms of labour organisation

The restructuring of the worldwide production of oils and fats, just like the reorganisation in the production of carbohydrates and proteins, is based on an increased interchangeability of product groups and producers. This restructuring process takes place along three general lines, each of which reinforces the other.

Horizontal interchangeability of vegetable oil sources

More than eleven oil sources can be divided into four groups, and within each group the producers of the different oil crops can be played off against each other (see Table 10.1).

The application of new plant-breeding techniques and enzyme technology directed towards attaining higher yields and modifying the fatty acid composition of different vegetable sources will intensify competition within the four major groups of vegetable oil sources.

In the group of *triaglycerols* the competition between rapeseed and soya might be intensified, which in turn could lead to a greater supply of linolenic acid. Also within the second group, the *diglycerols*, different biotechnology developments might lead to an increased level of competition between farmers in different oil producing crops. Almost an identical amount of linoleic acid is present in arachide, corn, soya and sunflower, while palm oil contains only a very small amount of linoleic acid. Plant breeding techniques applied to annual crops like corn, soya and sunflower might increase the production of linoleic acid from these crops. The increased competition within the diglycerols might lead to price decreases which the farmers in developing countries, especially the producers of arachide oil in Senegal will find difficult to bear.

Competition within the group of the *monoglycerols* is centred on the production of oleic acid. This acid can be extracted from arachide oil produced in, for example, Senegal or United States corn oil, Malaysian and Indonesian palm and palm kernel oil, EC sunflower oil and, to a lesser degree, Brazilian soya oil. Each modification in one of these regional production systems will have an impact on the world oleic acid production. Changes might come rather quickly because biotechnology will tend to speed up plant-breeding techniques in annual crops like corn, soya and sunflower. Therefore, the relative advantage of arachide oil, with the highest amount of oleic acid, might be reduced very soon.

In addition to the competition between vegetable oils and animal fats for general applications, coconut oil and palm kernel oil are the two most important sources for the supply of lauric and myristic acid within the group of the saturated oils. With an increasing production of palm oil the supply of palm kernel will also increase. Because of its residue character, such competition could hardly be borne by the coconut farmers, especially those operating small-farming systems.

Vertical interchangeability of vegetable oil

Biotechnology will also speed up the breeding of new varieties and the development of better extraction techniques. Therefore, it will become easier and cheaper to get the fatty acids from a broader spectrum of sources. Moreover, enzyme technology, with the possibilities of splitting off, adding to, and transferring certain fatty acids from different sources, could eliminate the differences between the tria-, di-, monoglycerols and saturated oils and fats. The interchangeability of oil sources will no longer be limited to sources with an identical composition but will be extended to oil sources in general. The specific character and advantage of the different oil sources based on their unique composition of fatty acids will disappear.

The possibility of breaking through the frontiers of the four oil groups will intensify competition between the oil sources in horizontal as well as in vertical direction. The different sources will increasingly become bulk products in which only price differences determine which source will be used as the raw material.

The blurring of sectors

The differences between the various vegetable oil sources might also disappear because of the possibility of adding fatty acids, produced in the factory by micro-organisms. By using these acids as additives flexibility in the supply of vegetable oil sources will increase. In addition to the development of new, or more efficient lipases the interchangeability of farmers cultivating different oil producing crops might be reinforced by the microbial production of food components. The world food system could be characterised by a new and sharper international control of farm and industrial labourers from different agricultural and industrial branches, unless farmers' and workers' organisations come up with new answers to the 'crisis of the sector' and adapt their joint action and structures to this new food production system. Moreover, relations between the food-components assembling companies and national political institutes will also change.

Enzyme manufacturers as 'political agents'

The blurring of existing differences among agricultural products (for instance between sugar and corn, between palm oil with stearic acid and cacao butter) and of sectors (between agriculture and biochemical production of acids) leads to new political relations within the international food system. The companies developing new catalysts (enzymes) and microbial production processes are playing an increasingly important role. They have at their disposal the scientific and technical capacities for making a growing number of farm crops and industrially manufactured food components which are mutually interchangeable. This means that an increasing number of farmers and workers from the different sectors in the different countries can be played off against each other.

The more that farming products are regarded simply as the sum of biochemical food components (carbohydrates, proteins and fats), the more the food industry can start procuring these components from industrial sectors too, as is evident from the biochemical production of proteins, amino and fatty acids. At the same time farming products can be used as inputs for non-nutritional ends. The segregation of the farming product and its intrinsic quality as a food product partly places the farming sector outside the social organisation of the agro-industrial chain of production. Consequently, the organisation of labour in the farming sector is no longer determined merely by the relationships the farmers have with the food industry but also by the developments that in part take place outside the food chain. The development and control of new catalysts such as enzymes will increasingly become vital in defining the position of certain agricultural branches within the food system. Enzyme development will result in a changed international agricultural product flow and this will

influence the price level of subsequent agricultural products. The introduction of biotechnology derives its great political significance from the
fact that research and applications concentrate on the strategic links in the
agro-industrial chain of production, such as seed for sowing, basic chemical
ingredients, enzymes, amino and fatty acids. These goods can be labelled
as 'politicising products' because they fulfil an important role in the
worldwide restructuring of farming and food production. 'Long-distance
supervision of agricultural production' and 'mutual interchangeability of
product groups and their producers' are central concepts for determining
the changes that are taking place. Attempts to privatise the crucial links
in the restructuring process of worldwide food production are in keeping
with these structural changes. Should companies be able to privatise these
'politicising products', they would gain an exclusive right to effect certain
political changes in worldwide food production. In this way they would
pre-eminently become the political agents, while the producers would no
longer be able to exercise control over the development and use of the
crucial links of the food chain.

Regional self-reliance

The improved production of catalysts will also increase the opportunities
in different regions to assemble the necessary food components. Enzyme
applications and microbial production will make possible the manufacturing of various food components from locally available agricultural
products. Next to an increasing flexibility in food production and an
opening up of new trade channels, an apparently increased autonomy
of regional food supply might arise. Even non-nutrient sources could
be converted into food components. Therefore, food production might
be organised on a more regional level, using techniques to process local
sources into the desired combinations of food components. Herewith, not
only local vegetable oil sources or animal fats might be processed into the
desired composition of fatty acids. Also carbohydrates might be used as a
raw material for local oil production. It could become possible to convert
glucose (from several carbohydrates) into certain fatty acids and put these
acids together with the cheapest vegetable oil sources of the world market.
Together with an increasing use of new scientific and technical extraction
methods, a more intensive use of local sources might be realised. Food
components might increasingly be extracted from a broad spectrum of
local sources.

Regionally self-sufficient food systems and increasing dependence on
the scientific and technical extraction techniques will become the structural

elements of the new organisation of the world food production system. In this new system the relation between local and international economic developments might be altered.

The several regional production methods by which the food components can be delivered from different agricultural and biochemical sources might become interchangeable production targets for the food conglomerates assembling the food components at an international level. The international competition will no longer be simply competition between different vegetable oil sources (products) but between regional methods of producing and converting food components.

Competition could even arise between the producers (and farmers) in different production chains (the energy chain with the conversion of ethanol into food and the agro-industrial chain with the conversion of agricultural products into food).

This changeover to a new production system, based on a flexible supply of raw materials converted into food components at the regional level, will modify the position of several regional farming systems within the world food system. This reorganisation of the position of regional farming systems will be dealt with in the next section. The restructuring of palm oil production is the context in which the new position of farming systems will be described.

IMPLICATIONS FOR PRODUCTION STRUCTURES

Three basic systems can be distinguished in palm oil production: (a) production by smallholders collecting from semi-wild oil palms; (b) production by smallholders, cultivating improved oil palms in pure stands; and (c) production on estates (Moll, 1987).

Within the above three production systems, a further distinction can be made between estates owned by private sector individually or in groups of people and estates owned by state or parastatal organisations and, further, between smallholders with improved palms who are integrated into an organisation providing comprehensive services and those who are independent. The result of this sub-division is five types of producers using three production systems (Moll, 1987). The areas cultivated by each type of producer for several producing countries are given in Table 10.4 as far as statistical categories used in the respective countries allow.

The socioeconomic position of these different sub-systems of producing palm oil might be altered by the introduction of clonal palm plants. The application of clonal propagation techniques seems to speed up the

TABLE 10.4 *Production area and production system in palm oil producing countries (1982, 1000 ha)*

Country	Estates		Smallholders		semi-wild palms
	state	private	improved palms integrated	independent	
Indonesia	259	101	6	–	–
Malaysia	–	646	580	–	–
Cameroon	37	14	–	3	100
Côte d'Ivoire	52	11	38	–	100
Nigeria	70		85	–	1650
Sierra Leone	9	2	5	–	500
Colombia	–	50	–	–	–
Honduras	–	6	14	–	–

SOURCE: Moll 1987, 74.

cultivation of high-yielding palm oil varieties on large-scale plantations.[2] As a consequence the traditional system, based on the harvesting of semi-wild palms might be squeezed out, while plantation production will increase.

Shifts to palm oil plantations

Palm oil production on large-scale farming systems is expanding. Unilever is introducing cloned oil palms throughout the tropical world – from Colombia to Brazil, from West Africa to Indonesia, Malaysia and the Philippines (RAFI, 1988, p. 102). As of 1985 the company had approximately 66000 hectares devoted to oil-palm plantations in Malaysia, Colombia, Ghana, Zaire, and Thailand (RAFI, 1988).

In Malaysia, cloned oil palms have reportedly increased yields by 30 per cent (RAFI, 1988, p. 118). Malaysian palm oil exports in 1985–86 exceeded 4.8 million tons. They now constitute about one-quarter of vegetable and marine oils traded in international markets. Almost half of the world's increase in edible oil trade during the decade of the 1980s was due to increased exports of Malaysian palm oil (RAFI, 1988). Despite the impressive yields realised in recent years, the oil palm boom has not benefited Malaysian producers. Because of enormous surpluses they are being compelled to sell below the cost of production. Virtually all Malaysian palm oil manufacturers produced at a loss in 1986. Not only in South-east Asia but also in Central and South America an increase of large-scale palm-oil production is taking place.

Palm oil production in Latin America has more than doubled since 1979, with Colombia, Ecuador and Costa Rica accounting for almost 80 per cent of the region's total production. In Ecuador, Latin America's second largest producer of palm oil, more than 20000 hectares were recently cleared to establish oil-palm plantations in the Amazonian jungle region. National and international companies with governmental support try to plough the jungle and regard the indigenous people as an obstacle in the way of progress. According to the leaders of the 'Federation of Indigenous Peoples of the Ecuadorian Amazon', the cultivation of oil palm in the region threatens the lives of 115000 indigenous people (RAFI, 1988).

However, not only multinationals like Unilever and United Fruit Company[3] are experimenting with clonally-propagated oil palms for large-scale production, but other actors have also initiated the same kind of research programmes, ranging from multinational corporations to government institutions in developing countries, from small biotechnology companies to large plantation groups.[4] This restructuring process, which leads to an increasingly flexible supply of vegetable oils to the food-processing industries, will also affect the economic position of producers of competing vegetable sources, like the coconut and arachide oil producers in Third World countries (see Table 10.1). The small-scale farming systems producing these competing oil sources might be especially hard hit.

Shift to large-scale production systems

The high palm oil yields will also affect the production systems of other vegetable sources. Viewing the quasi-identical composition of fatty acids the producers of arachide oils in Senegal, for example, may be hard pressed by the lower prices and greater supplies of the oleic acid from palm oil.

The increasing palm oil production will also strengthen the competitive position of its residue product, palm-kernel oil. Palm kernel might be used as a substitution for coconut oil on the market for saturated oils (see Table 10.1).

Countries like the Philippines, where coconut products are the most important export item, will be especially hard hit. In the Philippines, some 700000 small farmers grow coconuts on plots averaging less than 5 hectares. Nearly one third of the Filipino population is dependent on the coconut industry. A worldwide glut of low-priced palm oil will undoubtedly lower the prices of competing oils and could cause severe displacement of Philippino coconut producers with a longlasting impact on the Philipine economy.

The worldwide flow of cheap palm oil and other edible oils might also lead to replacements in cacao production. If the applications of biotechnology become successful in developing cacao butter substitutes from cheap palm oil and other edible oil sources, these oils might capture a large share of the cacao butter market. A shift could take place from raw material production for cacao in West Africa to palm oil in Malaysia, Indonesia and Brazil (Svarstad, 1988).

Half of the world's cacao crop is produced on small land holdings. Just seven countries – Côte d'Ivoire, Ghana, Brazil, Cameroon, Nigeria, Malaysia and Ecuador – account for 80 per cent of the total world production. Africa accounts for 57 per cent of world production, Central and South America account for 34 per cent, and East Asia for 9 per cent (RAFI, 1988).

The position of the cacao producers in these countries might be strongly affected by inter-esterification techniques. In the struggle to maintain a position on the cacao market, farmers are stimulated to cultivate new cacao varieties with higher yields.

Indeed, researchers of the Pennsylvania State University in the United States hope to develop such new varieties which will yield at least 1000 lbs per acre. Presently, the average yield of cacao producers is 350 to 400 lbs of beans per acre, worldwide. It is likely, however, that the benefits of high-yielding cacao varieties will be skewed towards the large-scale cacao plantations. To realise the potentials of the new high-yielding varieties, high standard cultivation is necessary. Plantations can do that, not small farmers (Svarstad, 1988).

Small-scale cacao producers in Africa, where the majority of the world's cacao is now produced, will be at a particular disadvantage. Brazil and Malaysia have a higher proportion of large size plantations and are able to apply advanced technologies quickly. It is probable that African growers will be hard-pressed to achieve the high production levels of Brazil and Malaysia in the near future. Malaysia is already the fastest growing cacao producer in the world. Malaysian cacao production increased tenfold between 1974 and 1984, and an estimated 625000 acres of cacao will be in production by the year 2000 (RAFI, 1988, 113).

The possibility of replacing cacao beans by other oil sources as raw material for cacao butter production, if realised, could lead to a shift in cacaobean cultivation by the small-sacle cacao holdings to the large-scale palm oil plantations. Also a relocation might take place from the small cacao production in West Africa to the large-scale cacao production in South-East Asia, especially Malaysia. In general, biotechnology may facilitate a shift in world agricultural production from small-scale to large-scale production systems. However, the economic necessity of real-ising an increasingly flexible supply of raw materials will undermine itself if production is concentrated too strongly. Therefore, biotechniques might also be used to speed up the integration of small-farming systems.

The integration of small-farming systems

In the coming period many initiatives might come up to extend the dif-fusion of plant breeding techniques, enzyme techniques and the microbial production and conversion of food components to small-scale farming systems. An increased intensification of the international competition between several regional methods of cultivating vegetable oils will be the logical consequence. For example, the distribution of highly productive coconut clones to the small farmers possessing less than 5 hectares could increase yields five times higher than in non-selected varieties (Sasson, 1988).[5] The competition between coconut oil and palmkernel oil will be intensified, greater supplies and lower prices will be the outcome.

The several possibilities of extending biotechnological applications to small-scale farming systems could be described as an appropriate biotechnological research programme. However, the increasing possibilities of integrating small-scale farming systems into the international food system should not be interpreted as a means of escaping the interchangeability tendency, or as a programme directed at an increased economic and political autonomy of small-scale farming systems. On the contrary, by speeding up the integration of the small-scale farming systems into the dominant relations of production biotechnology might increase the 'political marginalisation' of these interchangeable production units. The small-farming systems become increasingly dependent on international food companies, collecting food components produced in several regions, as well as on the services from companies supplying the new means of production, like seeds and information.

There has been a long history in industrialised countries of reducing the food problem to a technical or an economic issue, whereby the general idea was that an increase. in productivity would solve the food problem. The extension of biotechnology to small-scale farming systems might follow that tradition. Not a technocratic utilisation of biotechnology, but most of all political measures should be taken to prevent an increasing dependency of farming systems on service companies or food processing companies. Otherwise not development, but a marginalisation of small-scale farming systems will be the outcome of the expansion of biotechnology.

CONCLUSION

This brief review clearly demonstrates that the introduction of biotechnology in the vegetable oil industry will lead to two principal types of effects: (a) substitutions among vegetative sources of oil (because of reduced differences in their industrially useful chemical composition) and, (b) a restructuring of its production which would exclude the small farm sector and concentrate production in the large-scale units and plantations. Both will affect developing country exports, geographical distribution of production, the international division of labour, labour-utilisation in both crop production and oil extraction and the distribution of gains in output and incomes in each segment of the vegetable oil industry. The study also reveals the need for further research to quantitatively assess the magnitude of the impact on employment, the pattern of the international division of labour and income distribution through empirical case studies on the application of the various types of biotechnologies in this industry.

NOTES

1. In addition to the competition between the different oil sources on the world market there are also large countries, like China and the USSR, that are self-sufficient in oils (Tanis, 1987).

2. The research cooperation between Harrison & Crossfield and Unilever, one of the world's largest food corporations and the largest buyer and seller of oils and fats – in 1986 annual sales of more than $25 billion (RAFI, 1988) led in the 1970s to the successful application of clonal propagation of oil palms by tissue culture. Despite certain setbacks related to problems of transferring clones from the labs to the fields, research on clonal propagation continues and the first field planting of clonal plants was made in January, 1977, at Unilever's Palmol plantation in Malaysia.

3. United Fruit Company has entered into an agreement with a small United States biotechnology company, Agrogene Plant Sciences, which is developing cloning material for United Brands palm oil operations throughout Central-America (RAFI, 1988).

4. For example, multinationals like Unilever and United Fruit Company (a subsidiary of United Brands) are conducting research on clonal propagation of oil palms. The Palm-oil Research Institute of Malaysia (PORIM), a government sponsored institute which supports the nation's palm oil industry, is conducting its own tissue culture research in collaboration with plantation groups. Commercial plantings of cloned palms are foreseen for the 1990s. In Kasragod, India, the Central Plantation Crop Research Institute has reportedly developed its own technique for producing clonal plantlets of oil palm. A small United States biotechnology company, Escagen Inc., has an agreement with Malaysia's largest oil palm company, Sime Darby, on the development of clonal propagation techniques (RAFI, 1988).

5. However, as noted in Chapter 4, there is still no available technique for the efficient micropropagation of coconuts.

11 Policy Perspectives and the Future Outlook

Iftikhar Ahmed

INTRODUCTION

In this concluding chapter an attempt is made to highlight and synthesise the major issues raised by the individual chapters. The overview of the dominant themes dwells on the following. Firstly, it deals with output gains from biotechnology applications. With respect to employment generation the aspects such as labour absorption are covered, including biotechnology's impact on the rural labour market. The scope for multiple cropping, scale neutrality of biotechnology at the user level and shifts in cost structure of farming are other important issues considered. The review analyses how biotechnology can be designed and deployed to fulfil critical socioeconomic needs. It examines the impediments to the Third World countries' access to socially beneficial transgenic plants and microbes, and the differential impact of a given biotechnology in different socioeconomic settings. Third World technological and institutional capacity with respect to labour-intensive 'second-generation' biotechnologies are critically appraised. The relevance of existing agrarian structures on social differentiation is analysed. Finally, the result of the restructuring of the economies of countries affected by external shocks is discussed. Conclusions are also drawn on the intersectoral repercussions of the resource-saving character of some of the biotechnologies.

More importantly, this chapter tries to reinforce the major findings of the individual chapters by furnishing additional empirical evidence and introducing more rigorous analytical arguments on each of the above issues to give them a greater global perspective. The basic approach of this chapter is to further develop the analytical arguments, reinforce the empirical evidence and refine the conceptualisations contained in the volume to bring out clearly the contribution made by this volume to theoretical work and policy analysis. Finally, the chapter identifies the research gaps left by the volume and the social science literature as a whole, and tries to chart some of the priority areas essential for policy making and improving the understanding of the socioeconomic implications of the newly emerging biotechnologies.

OUTPUT GAINS

Evidence from China reveals that gains in yield and incomes through the use of biofertilisers and biopesticides in a wide range of crops have been remarkable (Table 11.1). The use of one traditional biotechnology rice variety, 'Shan Yu 63', which can resist the rice blast disease, increased output by 4.7 million tonnes (valued at 1.9 billion yuan) in 1987 and saved 100 million yuan on chemical pesticides, producing a total return of 2 billion yuan. Some indication of profitability is given by the return of 15.7 yuan from 1 yuan of investment in biofertilisers for wheat. It is significant that breakthroughs in cellular engineering included important cereal crops (rice and wheat) in addition to potatoes, sugarcane, bananas and tobacco (Table 11.2). Again, the increases in yield have been remarkable. For instance, by applying chromosomal engineering the genetic material from Sui Yan wheat straw was used to create wheat varieties like 'Xiao Yan' Nos. 4, 5 and 6, which incorporated the capacity to resist drought, dry hot wind and plant diseases. These modern biotechnology wheat varieties have been planted in ten provinces of China along the lower and middle reaches of the Huang He River. The total area sown under this technology is 2 million hectares and the wheat yield has increased by 900 000 tonnes since (Table 11.2). Similarly, the total sown area of rice, wheat and tobacco produced by the pollen haploid breeding technique was 466 700 hectares between 1981 and 1985.

The adoption of the new biotechnology in the cultivation of potato in Kenya doubled land productivity, labour-intensity and profitability and increased labour productivity by 24 per cent (Table 11.3). Its contribution (value added) to national income is twice that of the traditional technology. The relative efficiency (value added as a proportion of gross output) is also higher for the adopters.

The application of nitrogen fixing biotechnologies in maize could increase per hectare yields by 0.5 tonnes, i.e. 26 per cent on 5.2 million hectares cultivated by small farm holdings in Mexico. While national output would increase by 21 per cent, Mexican farmers' income would increase by 55 per cent (Gilliland, 1988) assuming constant output prices. This evidence, though limited, provides hopes to the policy makers that the present yield barrier can be overcome by the biotechnology developments.

TABLE 11.1   *China: Output and profitability of biofertiliser applications in selected crops, c. 1987*[1]

Biofertiliser	Crop	Area (ha)	Increase in yield	Increase in revenue
Root-combined nitrogen-fixation bacteria	Wheat	150,000	76.2 million kg	51 million Yuan[a]
Nodule bacteria	Peanuts	60000	340 kg/hectare	10 million Yuan
Nodule bacteria	Lucerne	200000	20%	45 million Yuan
Fungus (gemma fungus)	Rice, wheat, corn, etc.	100000	10–30%	
Actinomycetes	Grain crops, tobacco leaf, etc.	60000	40–50 kg/hectare (for grain)[2]	

1.   Revenue increased by 15.7 yuan per 1 yuan investment in biofertilisers.
2.   Average increase of pesticide cost amounted to 1.8–6 yuan per hectare.
SOURCE: Yuanliang, 1989.

TABLE 11.2   *China: Output gains from cellular engineering on major crops*

Crop	Type of cellular engineering	Crop variety	Sown area (ha) in 1988	Increase in yield
Wheat	Chromosomal engineering	Nos. 4, 5, 6, Xiao Yan	2000000	900,000 tonnes
Wheat	Culture of pollen haploid cells[1]	No. 1 Jing Hua	70000	15–20%
Rice	Pollen haploid	Xin Xiu, Wan Gen 959, etc.	170000[2]	About 10%
Rice	Pollen haploid	Nos. 8, 9 Zhong Hua	70000c	15–20%
Rice	Marker rescue	No. 1 Hu Yu	3000c	15%
Potatoes	Tissue culture		70000[3]	Over 50%
Sugarcane	Tissue culture		4000	Over 50%
Tobacco	Pollen haploid		10000	Over 50%
Banana	Tissue culture		100000 test-tube seedlings	Over 50%

1. Pollen and ovules have half the number of chromosomes present in all other tissues of a plant. By chemical treatment this number can be doubled so that plants generated from these cells have two sets of identical chromosomes and therefore identical genes. These homozygous plants are very useful in searching for mutants and for breeding.
2. 1985.
3. 1984.

SOURCE: Yuanliang, 1989.

EMPLOYMENT GENERATION

An attempt is made below to summarise the limited case study findings on the direct labour absorption in agriculture, indirect employment created through the backward linkages to crop production, the structure and stability of employment and the impact of advanced plant biotechnologies on the rural labour market. This information would improve the policy makers' understanding of the wider employment potentials of the advanced plant biotechnologies.

Labour absorption in crop production

Evidence from several ILO country case studies clearly suggests that with the application of advanced biotechnologies there is a saving in labour-use for chemical means of plant protection. For instance, the case study for Mexico (Chapter 4) shows that the application of micropropagation techniques need not lead to labour displacement in citrus cultivation, as this would be compensated by more intensive labour-use in weeding, pruning, irrigation and harvesting from a reduction in crop losses (labour accounted for 78–82 per cent of total costs of citrus production). In fact, in Malawi and Kenya there was a substantial increase in labour-use per unit of land following the application of the new biotechnologies (through the introduction of new practices) which also increased yields. In Kenya the doubling of labour intensity per unit of land was due to more labour needed for ridging before cultivating potato, and in Malawi for nursery and planting operations (Tables 4.5, 6.8 and 6.9).

Seasonal employment

As a result of the application of APB to a range of crops in several case study areas, a high potential has been noted for an indirect and steady source of employment through a strengthening of the forward linkages to the juice processing plant (Mexico), poultry production (Nigeria), coffee, henequen, tequila and dairy industries (Mexico), and the tea industry (Kenya). Underemployment (e.g. 75–90 per cent) of the agricultural workers in the southeastern region of Mexico, caused by seasonality of agricultural production, can be reduced by applying APB to create and widen crop varieties which prolong the growing season and supply of ripe oranges. It not only cuts down agricultural underemployment but also reduces excess capacity in juice processing (e.g. greater utilisation of the orange juice processing plant which is idle for 6 months).

Structural composition of rural employment

There is a structural change in rural and agricultural employment associated with the use of APB. Application of APB in China releases labour from agriculture which is absorbed in new and sideline activities in specialised occupations with a change in social organisation of the delivery of these services. Increased labour use in the agriculture of Malawi and Kenya has been brought about through structural adjustments resulting from new farm practices associated with APB (Tables 6.8 and 6.9).

Indirect employment

The case study evidence from Kenya shows that the backward linkages to input suppliers are enhanced by the use of the new technology, as shown by a larger flow of intermediate inputs per hectare for the cloned potatoes than for traditional farming (Table 11.3).

In general, it can be concluded that the chemicals used in rural areas are not produced there. Over 40 per cent of the chemical fertilisers are imported by the developing countries from the industrialised countries. The production process for both nitrogen and phosphate requires extremely large-scale capital-intensive plants. In fact, chemical fertilisers are among the most capital-intensive products made by man (Johnston and Kilby, 1975). A large chemical plant can cost anywhere between US$300 and 700 million (Doyle, 1985). Reducing dependency on chemical fertiliser through genetic applications will not lead to any serious labour displacement in Third World countries. This is particularly true of Africa, where a negligible proportion (1.2 per cent) of the world's total nitrogen fertiliser is produced (Keya *et al.*, 1986). On the other hand, it will relieve the balance-of-payments situation in most developing countries, and even in those with their own fertiliser industry the elimination of the dependency on chemicals offers the opportunity to reallocate vast amounts of scarce resources to more labour-intensive sectors.

In addition to the implications for the above traditional types of backward linkages to agriculture, a new type of backward linkage is emerging through the blending of two distinct types of workers' skills. Case study evidence presented in this chapter and in Chapter 2 clearly demonstrates the blending of workers with 'low-tech' skills to engage in traditional agricultural work with 'highly skilled' technicians directly involved with the advanced biotechnologies. This was seen for the micropropagation of crops in Mexico (technicians and scientists in the laboratory and the greenhouse), which applied plantlets for cultivation by traditional agricultural

TABLE 11.3 *Biotechnology and farm size: Potato and tea in Kenya 1987*

Key Indicators	Potato farms (N = 33)			Tea farms/estates (N = 39)
	Biotechnology (BT)	Traditional technology (TT)	Relationship with farm size	Relationship with farm size (biotechnology only)
Labour productivity (gross output/ha in shillings)	33 210	16 382	Inverse for BT and TT	Inverse
Labour intensity (work-days/ha)	301	144	BT: unclear TT: inverse	Unclear
Labour productivity (kg/work-day)	124	100	Positive for both BT and TT	Positive (sh/work-day)
Labour's factor share (wages as % of value added)	27	23	Positive for both BT and TT	Positive
Capital use sh/w-day sh/ha	3 867	3 426	Inverse for both BT and TT	Positive Positive

TABLE 11.3 *Continued*

Key Indicators	Potato farms (N = 33)			Tea farms/estates (N = 39)
	Biotechnology (BT)	Traditional technology (TT)	Relationship with farm size	Relationship with farm size (biotechnology only)
Intermediate inputs (sh/ha)	3 553	3 008	Inverse for both BT and TT	Inverse
Value added as a proportion of gross output (%)	89	82	BT: positive TT: inverse	Positive
Profitability (gross output minus operating costs in sh/ha)	20816	9916	Inverse for both BT and TT	Positive
Income ratio[1]	8	4	—	3 (small farms[2]) 2 (large estates[3])

1. Ratio of income of the 30 per cent of richer farmers to income of the 70 per cent of poorer farmers.
2. Up to 3 ha.
3. Over 20 ha.

SOURCE: Calculated from data in Chapter 5.

labour. In addition to the employment created for the traditional workforce in agriculture, 933 people, mainly scientific personnel, were employed by the Tea Research Foundation of Kenya in 1986 generating an income of Ksh 704 371 (Chapter 5). About 512 plant scientists are engaged in cellular engineering in China. Similarly, a scientific personnel can produce 8000 to 10 000 potato plantlets per day through the application of micropropagation techniques in Nepal which can then be easily handled by semi-skilled workers for rooting them in sandbeds (Rajbhandari, 1988).

Another interesting feature noted both for the Philippines (Chapter 9) and Mexico (Eastmond, 1989) is the domination of women in the micropropagation laboratories. For instance, women constitute 80 per cent, 74 per cent and 85 per cent of the Philippine Society for Microbiology, Cell/Molecular Biology and Biotechnology Societies respectively. In both these countries, these were considered as lowly paid jobs concerned with basic science with previously limited linkage to industry. Moreover, the work in the tissue culture laboratories is tedious, requiring patience and perseverance.

As regards employment created (cutting down seasonal underemployment) through the forward linkages, it is clear from the case study chapters that many more jobs could be created in orange juice processing, poultry production and the coffee, tea, henequen, tequila and dairy industries in a range of Third World countries.

RURAL LABOUR MARKET

We first look at the data for hired wage labour from the case study on Kenya (Chapter 5) and then deduce implications for labour use, particularly for women workers on the basis of the Green Revolution experience.

Wage labour
Application of APB increased the demand for hired labour (e.g. in Mexico for citrus and in Kenyan tea and potato), boosted wages, improved labour's factor share and reduced rural-urban wage differentials. Gross earnings from APB in Kenya compare favourably with wage incomes in a modern sector job important for dampening the pace of rural-urban flow of income seekers.

Women's employment
Despite the availability of chemical herbicides, virtually the entire Green Revolution (GR) area relied heavily on manual labour for weeding. Several

characteristics of employment are important for an analysis of the social consequences of increased herbicide applications: (a) weeding is one of the most labour-intensive of all agricultural operations for GR crops; (b) the GR led to a significant increase in the demand for hired labour in weeding; for example, in Sri Lanka, hired labour use doubled (Hameed *et al.*, 1977); (c) overall labour use in weeding doubled or tripled over that of the pre-GR crops, for example, in Bangladesh (Ahmed, 1981) and the Philippines (Bartsch, 1977); (d) small farmers adopting GR technology recorded much higher labour intensity in weeding than larger farmers (Ahmed, 1981); and (e) women constituted between 72 and 82 per cent of the labour used for weeding (Unnevehar and Stanford, 1985). In the above context, it is clear that the introduction of genetically engineered plant varieties will lead to a substitution of chemical herbicides for manual weeding leading to a massive displacement of women labour. The trends indicate that not only will the genetically engineered biotechnology plant varieties introduce a new fixed cost for farmers by forcing them to purchase the herbicide genetically tied to the seed supplied by the same company, but it will also strike a colossal blow at the poor.

NEW HOPE FOR MULTIPLE CROPPING?

The single most important factor which contributed to greater labour use per hectare on the Green Revolution areas was the practice of multiple cropping facilitated by the early maturing varieties of cereals. The application of micropropagation techniques to potato could similarly help improve cropping intensity. Since potatoes in most Third World climates take only 40–90 days to grow (compared to 150 days in the temperate climates), it can easily be incorporated into the cropping patterns currently practised for cereals like wheat, rice and corn.

Thirty poor countries already have the capacity to micropropagate potatoes. It is a major source of food for the poor families in Africa, and some Asian countries like India, Sri Lanka and the Philippines. Indeed, micropropagation techniques have made potato the second biggest crop (by weight) after rice in Viet nam, and quadrupled the production in China over the past thirty years (*The Economist*, 13 October 1990). In Viet nam as well, micropropagation techniques have increased potato yields from 200 tonnes to 8000 tonnes per year on 450 hectares of land within a period of four years (1980–84) (Uyen and Zaag, 1985). These techniques have already brought about yield increases from 8 tonnes to 18 tonnes per hectare in Nepal (Rajbhandari, 1988). Micropropagation techniques for potato are attractive for employment creation and poverty alleviation for the following reasons: (a) year-round production of plantlets is possible; in one rural valley of Viet nam each family is able to produce up to 150 000 virus free plantlets per year (Walgate 1990); (b) saving costs and difficulties of physical transportation of potato tubers to the fields for planting; (c) by generating plantlets directly from tissues of the plant, a substantial (in the aggregate) volume of tubers spared from planting can now be eaten by the hungry; (d) disease-free planting material could significantly reduce production fluctuations from diseases, potato being particularly vulnerable to as many as 268 diseases and pests and indeed the late blight could wipe out more than 50 per cent of the total crop (Manandhar *et al.*, 1988), and (e) while yield increases benefit land owners, increases in cropping intensity would also benefit the landless by increasing the demand for hired labour.

There is good promise for genetic engineering breakthroughs for potato compared to most other plants, because of the relatively easier possibility of incorporating genes for disease and insect resistance.[1]

WILL SMALL FARMERS BENEFIT?

It appears that like the diffusion process observed in the case of the Green Revolution, the large farmers pioneer the adoption of APB in Kenya. However, economic inducements exist for all categories of farms for the diffusion of the advanced biotechnologies. For instance, the Chinese experience provides the proof of the profitability of APB, which is important for its diffusion. In Nigeria (Table 8.10) the high and escalating costs of vegetative sources of animal feed and the lower relative price of SCP could serve as an inducement (with a positive price elasticity of factor substitution).[2] More than half the small-scale growers of citrus in Mexico were willing to adopt APB-based disease-free planting material for their survival. APB is likely to be more *scale-neutral* than mechanical or the Green Revolution technologies. Although biotechnological innovations would strictly constitute variable costs, collecting information on which the decision to adopt the technology is based actually represents a *fixed* cost. This would constitute an important reason for a bias in favour of large farmers (Kinnucan *et al.*, 1989).

The Green Revolution experience shows that small farmers in south and south-east Asia adopted the technology only after the large ones had applied it and raised their yields. While the large farmers obtained 'innovator's rent', food prices had been pushed down by the time the poorer, later adopters were ready to sell (Lipton and Longhurst, 1989). Similar adoption patterns may be repeated for the BR technology in the absence of any institutional reforms (e.g. improving small farmers' access to resources), although relief from petro-chemical dependence may reduce this lag.

In Asia producers in non-GR areas, often the poorest, gained nothing from the GR. Indeed they lost when extra output from GR areas depressed the returns from their meagre output. For example, these farmers lost when the extra GR sales from the Punjab (wheat), or Central Luzon (rice), pulled down farm-gate prices in impoverished Madhya Pradesh (India) and Mindanao (Philippines) respectively (Lipton and Longhurst, 1989). Through its potential for less favoured areas, the BR may help redress this disparity. The same applies to African farmers, virtually untouched by the GR. The deductive reasoning used in Chapter 4 clearly demonstrates that biotechnologies could redress regional socioeconomic disparities in Mexico.

As we shall see from the following sections, the two most important contributions biotechnologies can make is to reduce resource-poor small farmers' cost of production and to reduce the risk-averse small farmers' variability of output due to environmental stress (disease, pest, drought and salinity).

SHIFTS IN FARMING COST STRUCTURES

Biotechnology developments could significantly shift production cost structures, leading to a reduction in farmers' average cost of production. Chemical nitrogen fertilisers account for 75 per cent of agricultural production costs in Brazil (InterAmerican Development Bank, 1988) and 60 per cent of the energy costs of wheat production in India (Jain, 1985). Fertilisers and pesticides constitute over 83 and 80 per cent of the costs of production of the Green Revolution rice varieties in Thailand (Palmer, 1976) and wheat in Europe (*The Economist*, 1987).

Exploiting and extending the domestic potential of Third World capabilities in supplying the 'second generation' biotechnologies is certainly a more attractive economic proposition as compared to import-reliance. Indeed, it is remarkable that commercially adopted micropropagation techniques in many developing countries are often cheaper. For instance, in Mexico disease-free stocks of flowering plants can be produced at half the cost of the imported supplies even after securing 60 per cent profit from its production. Similar comparative advantages are noted for micropropagation techniques for tequilina (Eastmond *et al.*, 1989). Furthermore, supplies of micropropagated planting material are not adeuqate to meet domestic demand. In the Philippines, commercial micropropagation of orchids is not sufficient for local needs, with imports continuing despite the fact that prices charged locally for the plantlets are still too high for the small farm sector (Zamora and Barba, 1990).

Similarly in Nepal, the use of imported certified potato seed leads to 40 to 60 per cent higher production cost and domestic micropropagation drastically reduces production costs of potato plantlets (Manandhar, 1988). Modifications to scientific procedures associated with micropropagation techniques reduce costs of potato plantlets from Nepalese Rupees 1.5 to about Rupees 0.30 – 0.50 per plantlet (Rajbhandari, 1988).

Biotechnology's capacity to protect crops from insect damage is more economic compared to alternative chemical means. For instance, damage by one single insect accounts for 40 per cent of all soybean crop losses caused by insects in Brazil. A virus introduced on 11000 hectares during 1983–84 led to about 75 per cent savings in the cost of protecting soybean crops as compared to that of chemicals (Inter American Development Bank, 1988).

At the national level, biotechnology could help save scarce foreign exchange. Imported chemical fertilisers (whose price is increasing) affect the balance of payments of Third World countries. For instance, India's bill for fertiliser imports from the late 1960s to 1980s rose by more than 600 per cent, an amount greater than the country's expenditure on food imports in the deficit years (Doyle, 1985). The magnitude of foreign exchange leakage is dramatically brought out by evidence from Nicaragua. Of every US$3 earned from export of cotton by Nicaragua, US$1 was returned to the United States to pay for pesticides (*New Scientist*, 14 July 1988).

DESIGNING BIOTECHNOLOGIES TO MEET SOCIOECONOMIC NEEDS

This volume provides sizeable evidence on how exactly biotechnologies ought to be designed by scientists to launch a planned assault to solve location-specific socioeconomic problems. These are highlighted below, first by defining and describing the socioeconomic problem faced by the poorest groups (producers and workers), and then the appropriate biotechnology solution offered.

In Chapter 4, it was noted that the orange leaf rust disease ravaging coffee cultivation in Mexico has threatened the survival of the over-whelming majority of the growers, who are small. Chemical means of control is beyond their reach. Application of APB to supply disease-free or – resistant plant materials will not only save but expand the employment opportunities of these producers and of the large body of hired labour, and generate indirect employment through its forward linkage to the coffee industry and backward linkages to micropropagation laboratories and nurseries to produce the plantlets.

Another illustration (not based on the above case study) concerns the *tequila* industry. The crop *tequila agave* used to produce the tequila drink is grown in the Mexican state of Jalisco. The plant takes nine years to grow to a mature stage before it can be utilised. Not less than 12 million plantlets are required to replenish existing stocks.[3] Micropropagation techniques again offer the answer. Some 6000 small contract growers who supply the large agro-industrial companies with the raw material will stand to benefit (Eastmond *et al.*, 1989). Through the forward linkages to the tequila industry, additional and more stable employment will be generated.

SCP to combat protein malnutrition

One other example of the practical application of bitoechnology for social benefit concerns the animal-feed biotechnology. The Nigerian case study (Chapter 8) has clearly demonstrated that the application of SCP biotechnology could help alleviate protein malnutrition in general and boost animal protein intake by Nigeria's protein-deficient population.[4] Fortunately, the current aversion to red meat and red meat products on the on grounds of health underpins the relevance of SCP biotechnology to the poultry industry. The economic climate is favourable for its acceptance as can be seen from the following: (a) the income elasticity of demand for poultry products is higher than that for beef; (b) the supply-demand projections reveal an excess demand for poultry products; (c) relative prices of other sources of poultry feed (soybean and fish meal) compared to SCP are higher and on the increase; and (d) the ban on the import of poultry products and poultry feed provides the protection and opportunity for import substitution.

Prospects for SCP technology application is also bright for the Latin American region. Between 1979 and 1987 the import of soybeans and protein meal as sources of animal feed increased by 433 per cent to 516 per cent in Venezuela (Martel, 1991). The abundant supply of natural gas in Venezuela could easily be used to produce SCP as a major import-substitution measure. Indeed, in the Latin American region, Cuba has already established 13 plants for the production of SCP based on cane-molasses, a byproduct of the sugar industry (Inter American Development Bank, 1988; also discussed in Chapter 3).

DIFFERENTIAL IMPACT OF BIOTECHNOLOGY

In Chapter 7 it was demonstrated that the BST technology being developed and applied in the industrialised countries clearly has a differential impact in the Third World and industrialised countries. As we shall see below, in all of these countries it concerns both the dairy producers and the consumers of milk. In the case of industrialised countries, the developments could affect the structure of manufacturing BST as well.

Meeting the Third World's socioeconomic needs

Third World countries (and their small dairy farmers) could indeed stand to gain if the BST technology could be made available to them at the low price of US$1 per day per cow. As we shall see later, it can only be supplied so cheaply if the scale of production is large enough even if it meant that the tendencies of monopolistic production structures of BST would be reinforced in the industrialised countries as a consequence.

Advanced biotechnologies could play an important role in boosting the nutritional levels of the poor. As was noted in Chapter 7, in Mexico the BST technology could reduce the daily deficit of 12.5 million litres of milk and make it more accessible to the population (37 per cent of the population currently consume only 14.5 per cent of the available milk supply). It increases milk production in dairy cows by 10–25 per cent. This is like having extra milk without extra feed. Purchasing power could be increased by stimulating employment in the production and processing of milk and the feed industry, all of which are concentrated in a few hands. BST also holds prospects for Pakistan. Despite having three and a half times as much pasture as Wisconsin and one and a half times as many dairy cows, Pakistan produces only a quarter as much milk. Pakistan's cows are only 15 per cent as efficient as Wisconsin's. As a consequence, Pakistan has to spend about US$30 million importing (mostly dried) milk each year (*The Economist*, 13 January 1990).

It has been estimated that returns over variable cost could be as much as 26 per cent for dairy farmers using BST (Chapter 7). With certain assumptions of milk prices and costs, if BST induced a milk production increase of 15 per cent, a farmer with 500 cows could make an extra US$82000 profit per year in the United States (*New Scientist*, 24 March 1988).

Controversy in the industrialised countries

The opposition to the application of BST in the industrialised countries has been basically on two grounds: (a) consumer safety, and (b) BST induced concentration in the dairy industry (often described as the demise of the family farm).[5] The concern with consumer preference was so great that five of the largest US American supermarket chains announced that the milk and dairy products sold by their stores would not contain milk from cows treated with BST. This was alarming, as these included the largest American supermarket chain, Kroger of Cincinnati, which has 1200 stores in 30 states and Safeway Inc. which has 1100 supermarkets in the West and around the district of Columbia (Schneider, 1989). However, the United States Food and Drug Administration has announced the findings of its study which shows that BST treated milk production in cows is harmless to human health (*International Herald Tribune*, 25–26 August 1990). This has also been confirmed by a team of American doctors who announced that the BST causes no changes in milk composition of any practical importance to consumers (*Chicago Tribune*, 22 August 1990).

As regards the bias in favour of larger dairy farms, it is argued by four major manufacturers of BST that its cost of less than US$1 a day per cow would make it *scale-neutral* (Schneider, 1985). Moreover, others argue that the trend towards fewer and larger dairy farms (e.g. 30 per cent decline in the number of dairy farms in the United States over a short period of time) was already in existence irrespective of the introduction of BST (Buttel and Geisler, 1989).

However, to be able to offer BST at such low prices to farmers requires significant economies of scale to reduce the unit cost of producing BST. It is clearly demonstrated that increasing the scale of production of BST from 0.5 million to 7 million doses per day, reduces average costs from US$4.23 per gram to US$1.97 per gram (Kalter *et al.* 1984). If the size of the market were limited domestically to the United States, a plant with the capacity to produce 7 million doses would cover nearly two-thirds of the American dairy herd (Molnar and Kinnucan, 1989). Even lower unit costs (to the level of US$1 per day per cow as claimed above) through larger economies of scale can be achieved, as there is an even bigger global (approaching US$1 billion annually) and international (US$100–$500 million annually) market emerging (Schneider, *op. cit.*, and UNDP, 1989).

Therefore, there is an economic incentive to manufacture BST under monopoly conditions. It is hardly surprising that four giant multinationals, Monsanto, Eli Lilly, Upjohn and American Cyanamid are currently engaged in the development of BST (Schneider, op. cit.). While this offers the prospects of supplying BST cheaply to dairy farmers world wide including the Third World (making it more scale-neutral at the user level), it could create a monopsonistic market structure in the United States with concentrations both in the dairy and BST manufacturing industries.

GENETIC ENGINEERING FOR JOB CREATION

Table A.1 assembles the fragmented and widely scattered information on primarily single gene-based genetic engineering breakthroughs. This represents only the tip of the iceberg. The effort made here to harness and assemble information on all the activities that exist world-wide for the creation of transgenic plants and microbes was made more difficult by the fact that much of the work on genetic engineering is shrouded in secrecy.

Certain trends are clear from Table A.1. The private sector corporations dominate genetic engineering research, and their eyes are on agronomic traits and on crops which promote their markets for seeds and/or agrochemicals. These also concern crops of importance to industrialised countries as it is difficult to police patency infringements in Third World countries. Developing countries generally do not have patent laws. The private industry cannot recover revenue through royalties and licenses.

Private-public sector balance
Usually, the private and public sectors shared the total research expenditures in agriculture in the United States almost equally ($2.1 billion spent by the private sector as against US$1.9 billion by the public sector). As soon as biotechnology appeared on the research agenda, the agricultural R&D expenditures on biotechnology by the private sector ($150 million) was 50 per cent higher than that of the public sector ($95 million) in the United States. The bulk of the overall private sector expenditures on biotechnology was on human health care (61 per cent of the total private biotechnology research expenditures). Agriculture accounted for 23 per cent and others (mainly chemicals) accounted for the remaining 16 per cent (Farrington, 1989). The pharmaceutical sector is favoured by industry because there is a clearly defined path from laboratory experimentation to the market.

Certain features of the phenomenon noted are: (1) Private sector accounts for two thirds of the total global funding (US$4 billion) of biotechnology research in the industrialised countries, and (2) large chemical multinational companies spent 50 per cent of the total R & D budget on biotechnology. It is little wonder that they spent US$10 billion over the last decade to buy up seed companies to facilitate the marketing of their biotechnology products (James and Persley, 1989). Indeed after the year 2001, perhaps 75 per cent of all major seed will be based on genetic engineering or derived via tissue culture (McGrawhill Bitechnology, 1989). As a consequence, the cost of seeds as a proportion of total cost of production of wheat in Europe could be pushed up from the current levels of 20 per cent to 50 per cent at about that time (*The Economist*, 1987).

From the perspective of labour absorption and poverty alleviation, the following observations can be made on the genetic engineering break-throughs described in Table A.1 in respect of the *transgenic* (containing a foreign gene) plants and microbes:

(a) Pest and disease resistance and drought tolerance will reduce output variance, which is important for risk-averse farmers; while together with the breakthroughs for nitrogen fixation these will obviously reduce resource-poor farmers' costs of production, the magnitude of which was described in the section on shifting cost structures; further research by Cornell University Boyce Thompson Institute for Plant Research, which has discovered a bacterium that can fix nitrogen without depending on the plants for energy, should be encouraged (*Genetic Engineering News*, 1989);

(b) production of 'less thirsty' crops will increase labour absorption through area expansion and multiple cropping now made feasible;

(c) lower labour requirement in pest and disease control may be made up by overall increases in labour use in other new operations;

(d) it is quite clear that herbicide resistance will directly displace labour for weeding as was described earlier;

(e) prolonging the shelf life of freshly harvested agricultural produce will certainly help the poor confronted with inadequate marketing infrastructure (without access to preservation or refrigeration facilities in addition to facing communication constraints);

(f) genetic engineering breakthroughs in (a) and (b) above will also help compensate for the inadequacies of extension services and 'delivery failures' by, for instance, parastatal marketing boards responsible for supplying crucial crop inputs, typical of Africa;

(g) genetically engineered microbes may benefit the small farmers if

TABLE 11.4   *Comparison of chemicals and microbes for plant protection*

Criteria	Chemicals	Microbes
Costs/benefits		
R&D	US$20m	US$0.8–1.6m
Market size required for profit	US$40m/year to recoup development costs, therefore limited to major crops	Markets under U$1.6m may be profitable due to low development costs
Toxicological testing	US$10m	US$0.5m
Patentability	Well established	Still developing
Lead time	6–7 years	3.5 years
Discovery	Screen 15000 compound to identify one product	Rational selection for specific target pests
Efficacy		
Kill	100%	Usually 90–95%
Speed of kill	Rapid	Can be slow
Spectrum of activity	Generally broad	Generally narrow
Resistance	Often develops	Only one known case
Type of action	Can be both preventive and curative	Generally only curative
Safety		
Operator safety	Chemicals can be hazardous	Low operator risk
Environmental impact	Many examples, e.g. accumulation in food chains	Few examples with use of indigenous micro-organisms
Residues	Interval before harvest often required	Crop can usually be harvested immediately after application

SOURCE: Bunders, 1990, p. 40, Table 4.2.

these spill over to the poor neighbours' plots and fix nitrogen there or protect the crops from pests and diseases there; a rough comparison (Table 11.4) of chemicals with microbial controls shows the clear economic and safety advantages, although costs need to be further reduced and effective means of dispersing the microbes need to be developed to make their use feasible for smallholders;

(h) the major obstacles to the Third World countries and the poor farmers' access to the above beneficial labour-using biotechnologies are the legal and financial barriers associated with the proprietary rights over these technologies through patents; moreover increasing research partnership between industry and the universities and research institutes tends to diminish Third World countries' access to science and technology produced in public institutions which until then was available freely as a public good;

(i) world production will be boosted by genetically engineering pasture crop for sheep; it will help wool producing Third World countries by reducing average costs, and by saving on grazing land (extra wool without extra pasture).

IMMEDIATE EMPLOYMENT GENERATION: 'SECOND GENERATION' BIOTECHNOLOGIES

The 'second generation' biotechnologies of which micropropagation is the major component are within the scientific and financial reach of Third World countries. It has been estimated that a fully equipped laboratory excluding land and buildings might cost US$250 000 (Lipton and Longhurst, 1989).

Micropropagation techniques can be immediately deployed to enhance rural employment. The capacity to generate micropropagated disease-free planting material in Mexico (tequilina) and Nepal (potato) has already been demonstrated to be cheaper than the imported planting material. Indeed, this technique is already applied to potato in 30 poor countries as explained in the section on multiple cropping. Singapore has the capacity (Plantek International) to provide disease-free coffee plantlets for large-scale plantings throughout South East Asia. Similarly, the government agency in Brazil (EMBRAPA) is able to produce coffee plantlets (*Biotechnology Development Monitor*, 1990). We have already noted the capacities existing in Malawi, Nepal, Viet nam and Kenya.

The scientific capacity that developing countries possess is being channelled mostly to cover non-food crops and to meet the needs of the commercial and large-farm sector. This is clear from the developments in the Philippines, Mexico and India, although these may create employment indirectly for the hired rural workers.

The Biotechnology Department of the Indian Ministry of Science and Technology has supported (University of Delhi) micropropagation techniques for bamboo (40 000 plantlets by 1989–90), oil palm (5000 plants by January 1990), coconut and natural rubber. Some commercialisation has been initiated or achieved for cardamum and bamboo (Mani, 1990). Similarly in the Philippines, out of the 28 commercial and semi-commercial tissue culture laboratories owned privately or by the government, 22 are devoted solely to orchid propagation (Table 11.5). Only five laboratories propagate food/fibre crops and two laboratories which are solely devoted to food crops are both government owned.

Similarly in Mexico (Table 11.6) the micropropagation enterprises are geared to market plantlets for nonfood crops. Certainly, crops (maize, beans and other cereals) which constitute the basic Mexican diet receive little attention. Maize is grown primarily by the peasants and the application of micropropagation techniques to maize is not profitable under current prices.

A special technique developed in Japan could enable the production of roughly 3 billion rice seedlings to be grown from a single seed in about 6 months. This opens up vast prospects for the densely populated and hungry masses of the major rice growing areas of the Third World. This technique saves on seeds (releasing more grains to feed the hungry) and plenty more seedlings to plant for the countless unemployed hands. In labour-scarce Japan robots are being sought to meet the intensive labour demands for root separation of seedlings grown by the new technique (UNIDO, 1989).

COPING WITH EXTERNAL SHOCKS

Third World countries may annually lose US\$10 billion of their export income due to the biotechnology-based product substitutions (Kumar, 1988), with serious repercussions on the international division of labour. Moreover, this decline in LDC sugar exports was a direct consequence of the sugar policy of some of the industrialised countries, particularly

TABLE 11.5 *Commercial and government micropropagation laboratories (The Philippines)*

Classification of laboratory	No.	Grouping according to crops cultured			
		Orchids only	Orchids and ornamentals	Orchids and food fibre	Food/fibre only
Government-run	7	3	1	1	2
Private	21	19	0	2	0
Total	28	22	1	3	2

SOURCE: Zamora and Barba, p. 51, Table 1.

TABLE 11.6　Commercial and government micropropagation activities: Mexico

Firm (private/public)	Species	Total annual production of micropropagated plants	Investment in tissue culture laboratory (US dollars)	Number of employees in laboratory	Market % national	international
Biogenetica Mexicana, S.A. de C.V. (private)	Gerbera, Gypsophyla Dieffenbachia, Caladium, Spatiphyllum	100,000	90,000 (1984)	4	100	–
El Rancho, La Joya (Nursery) (private)	Orchids	100,000	100,000 (1985)	2	20	80
Invernamex (Nursery) (private)	Gerbera, Gypsophyla, Strawberry, Raspberry Blackberry	600,000 (1989)	400,000	8	100	–
Viveros 'El Morro' (Nursery) (private)	Spatiphyllum, Singonium	300,000 (1988)	400,000	7	–	–
FIRA (Bank of Mexico) (public)	African violet, Gerbera, Chrisanthemum Strawberry	100,000 (1988)	45,000	7	100	–

SOURCE: Eastmond, et al. Table 3.

the European Communities, which is the world's largest sugar exporter to the market economy countries. For instance, while 90 per cent of the sugar internationally traded came from the developing countries in 1975, it declined to about 67 per cent in 1981 (Otero, 1990). Sugar imports by developed countries declined from 70 per cent to 57 per cent during the same period. World consumption of HFCS accounted for only 1 per cent of total sweeteners in 1975, but rose to 6 per cent in 1985 (Wald, 1989). Thirty-four different soft drink manufacturers in the United States have switched to the immobilised enzyme technology (HFCS). As a direct consequence, sugar exports from the Philippines declined from US$624 million in 1980 to US$246 million in 1984. In the Caribbean the decline of sugar exports to the United States was of nearly the same magnitude during this period. This was accompanied by a crash in sugar prices from US cents 63.20 per kg in 1980 to US cents 8.36 in 1985 (Panchamukhi and Kumar, 1988). It is not surprising that the livelihood of over 50 million workers engaged in the sugar industry, mostly in Third World countries, was affected by the decline in their exports (Panchamukhi and Kumar, 1988).

The imposition of import quotas on sugar during the 1980s made the United States domestic price of sugar three and a half to nearly eight times higher than that of international prices (Maskus, 1989). In addition, the development of the HFCS reduced production costs to such an extent that a shift from sugar to HFCS was more profitable. This liquid sweetener (HFCS) became more suitable for industrial application. It is little wonder that two major beverage companies, Pepsi Cola and Coca Cola, shifted from sugar to HFCS (Junne, 1991). This induced technological innovation is now likely to become permanent as the powerful corn grower and millers lobby (supplying 97 per cent of the United States market for corn sweeteners) would like to see a continuation of this trade blockade against Third World sugar.

Japan imports both corn and sugar but higher taxes were imposed on sugar, which induced the adoption of HFCS. This decision was also largely influenced by political considerations, as the production of HFCS in Japan would require large imports of corn from the United States, which in turn would help Japan reduce the huge trade surplus with the former country (Junne, 1991). The displacement of Third World exports of sugar to both the United States and Japan was, therefore, the direct consequence of distortions in free international trade flows.

Similar tendencies to substitute vanilla flavour by biotechnology substitutes (through plant tissue culture) threatens 70 000 small farmers in Madagascar, which could also lose US$50 million of its annual export earnings (Mushita, 1989). Comoros will similarly be affected by this substitution (Junne, 1991). In fact, a Californian biotechnology company, Escagenetics has already developed a new tissue culture method for producing vanilla plantlets on a commercial scale clearly intended to capture the lucrative annual flavouring market worth US$200 million (*New Scientist*, 1991). The significantly lower cost of tissue culture compared to the traditional vanilla extract makes it more profitable. Cacao, the second most important agricultural commodity in the Third World faces a similar threat of substitution by biotechnology products. Africa accounts for nearly 60 per cent of the world production of cacao.

Small cocoa producers in Cameroon, Ghana and Côte d'Ivoire will be affected by the biotechnology developments in the Swiss-based company, Nestlé (Hobbelink, 1989). The European Patent Office has already received two patent applications from Kao Corporation of Japan for genetically engineered enzymes for making cocoa butter substitutes (Svarstad, 1988).

It is reported that current biotechnology research in Germany is oriented to come up with a substitute for coffee (Otero, 1990). Third World countries account almost exclusively for world coffee exports (US$10 to 50 billion worth of coffee beans each year). Apart from the adverse impact on the balance of payments of the major coffee exporters, Colombia, Burundi, Uganda, Rwanda and Ethiopia, the livelihood and jobs of 500 000 small producers (average size 1 hectare) in Rwanda and another 650 000 in Indonesia would be threatened (*Biotechnology Development Monitor*, 1989) if the above substitution efforts succeed.

While SCP technology will be beneficial to Nigeria and Venezuela, it will affect exports from Brazil and other developing countries (as elaborated in Chapter 3).

Another oil crisis?

Apart from the job losses associated with the changes in the North-South trade flows, biotechnology applications to convert plant oils to produce structural lipids or tailored fats will affect the market shares of 11 vegetative oil crops traded by the Third World countries (as was elaborated in Chapter 10). Biotechnology will increase dramatically the market for castor, palm and groundnut oil, and reduce 8 others (Kumar, 1988), although less so for groundnut oil. For instance, while coconut is the source of only 2 per cent of the world's oils and fats market, the Philippines alone supplies 80 per cent of the coconut. Decline in Philippine coconut export will directly affect the 15 million Filipinos dependent on the coconut industry for their livelihood, the majority of whom are poorer than the rest of the farming population (as was emphasised in Chapter 9).

The end result of this South-South trade war could be that a fewer number of developing countries will be fulfilling the international needs of a smaller band of vegetable oils, the final effect of which will be an overall shrinkage in international trade in terms of monetary value rather than in volume.

Therefore, there is an urgent need to internally adjust production structures of the affected countries to redeploy the workers, particularly from the plantations and small farm sectors made redundant by the decline in the international demand for their exported products. In order to minimise the adverse effects of such external shocks, it is necessary to be vigilant about new biotechnology developments related to other important export crops so that necessary structural adjustment measures can be adopted in good time. Third World countries will also be compelled to look for alternative export markets.

AGRARIAN STRUCTURE

Table 11.3 furnishes concrete empirical evidence in support of the inverse relationship between farm size and productivity – both under the traditional and the new biotechnology for both potato and tea. In all these cases, the small farms relative to large ones, make a larger contribution to national income, extract higher levels of profit and demonstrate stronger backward linkages to agricultural input suppliers. This certainly constitutes important evidence to the policy-makers about the development potential of small farm strategies, both in a traditional agricultural setting and under dynamic conditions of technological change. Clearly, such a strategy would lead to output gains, and prevent worsening income inequality without sacrificing employment.

Evidence from Malawi and Kenya shows that with the application of APB, labour's factor share increases. This share is already high in Mexico (78–82 per cent production costs of citrus growers). However, the past trend of increasing social differentiation will be accentuated in both Mexico (as is revealed in Table 4.2 through limited mobility across occupational class structures) and Kenya (increase in the concentration ratio), unless essential agrarian reform programmes are adopted. It is clear that the advent of biotechnology into an agrarian system in which land is unequally distributed (the smallest 70 per cent of the Kenyan farms accounted for a meagre 10 per cent of the country's land area) tends to

214   *Biotechnology: A Hope or a Threat?*

reinforce the existing inequality (Table 11.3). In this respect the Kenyan experience is certainly a close parallel to the experience of inequality created by the Green Revolution in Asia although it is likely to be less intense.

AGENDA FOR FUTURE RESEARCH

This volume identifies major research gaps and provides the setting for launching further empirical and analytical work in the following major areas: (a) rural labour markets; (b) biotechnology diffusion; (c) structural adjustment, in response to external shocks and the new international division of labour; (d) linking existing biotechnologies to solving current concrete but location-specific socioeconomic problems; (e) Third World countries' access to socially beneficial transgenic plants and microbes; (f) enhancing Third World capacity in biotechnology, and (g) nonfarm employment.

Rural labour markets

It is difficult to predict the impact on real wages of hired labour due to the dual effects of biotechnology on the rural labour market.[6] On the one hand, it displaces labour for pest and disease control and, on the other, it increases labour use on account of new operations and increased cropping intensity or area expansion. Therefore in order to determine who gains and who loses it is important to investigate whether the workers made redundant in certain traditional agricultural operations possess the skills for the jobs created for the newer operations.

It is extremely important to identify the category of labour displaced by biotechnologies. If it displaces primarily family labour, the impact on the market for hired labour will be limited if workers withdraw themselves and become economically inactive. Instead, it is also possible that the displaced family labour may present themselves on the hired labour market. On the other hand, if biotechnology applications lead to the massive displacement of hired labour, it will have a significant depressing effect on the rural wage rates for the related agricultural operation. Do the workers displaced in one agricultural operation possess the skills to undertake the tasks newly created by biotechnology?

Impact on employment (different categories of workers) depends on the existing labour market structures. It is important to examine, for countries without well developed labour markets, what opportunities biotechnology offers for wage labour. While it was noted that APB increases the demand for hired labour, it is often the *adult* male labour who is hired. Even if it increases the income of a typically landless agricultural labour family, the non-availability of the adult male member of the household leads to greater work burden on women for unpaid household work. In this respect, it would resemble the effect of migration by male family members.

Biotechnology diffusion

Essentially research on biotechnology diffusion needs to cover two issues, (a) scale neutrality, and (b) the effect of structural adjustment measures on biotechnology diffusion.

There is a need to identify specific policies which enable the small farmers to adopt the new biotechnologies at about the same time as large farmers. Prospects for lowering average costs (i.e. cost saving) particularly the fixed cost element of biotechnologies need to be explored. The potentials for biotechnologies which counter fluctuations in crop output would also need to be studied. An attempt made in this volume to address the question of scale neutrality is only a speculative and partial effort which has to be followed up by many more empirical studies before definitive conclusions can be drawn.

The question of a more equitable land ownership distribution remains important in the future research agenda for agrarian reform from the perspectives of growth, efficiency, labour absorption and equity, both under a traditional agricultural setting and under dynamic conditions of technological change. In this connection, it is to be noted that the high labour absorption capacity of agriculture in East Asian countries, such as China, the Republic of Korea and Taiwan, China, was primarily due to the egalitarian distribution of land, usually associated with a low incidence of hired labour (ILO, 1988).

Secondly, studies need to be undertaken on the contribution of structural adjustment measures to biotechnology diffusion in general. Two components of structural adjustment which could favour biotechnology diffusion need investigation: (a) price liberalisations and removal of subsidies could make the Green Revolution technology inputs more expensive; elimination of such price distortions could be an inducement to biotechnology diffusion; (b) privatisation of the distribution and marketing of agricultural inputs currently entrusted to inefficient parastatals and state corporations could again make biotechnology more attractive to farmers, since these would affect the relative costs of biotechnology and the chemical-intensive GR technology.

Structural adjustment, external shocks and the new international division of labour

In countries threatened by massive unemployment brought about by declines in their major agricultural exports on account of biotechnology-induced import substitutions in the industrialised countries, it is necessary to assess the need for the restructuring of the economies of the affected countries to be able to better redeploy and retrain the workers made redundant as a consequence. Particular attention is to be given to formulating measures for coping with the minor 'oil shocks' affecting trading patterns of several developing countries critically dependent on vegetative oil exports. In short, future research has to provide insights to measures which would mitigate or avoid the negative effects of changing trade patterns on Third World employment.

Future research also has to look at the implication of GATT negotiations for the removal of artificial barriers to Third World exports by industrialised countries. Chapters 9, 10 and 11 have shown how those trade restrictions have stimulated biotechnologies which eventually introduce structural rigidities making the substitutions that have occurred irreversible.

Biotechnology research should be directed to solving specific local socioeconomic and technical problems, particularly those of the poor in developing countries, in a systematic and well planned manner. This requires a dual approach and special skill. The social scientist would need to have a thorough understanding of the scientific and technical capabilities of the newly emerging biotechnologies. The same social scientist should then relate these capabilities to solving the socioeconomic problems which have been studied. This volume has partially demonstrated the appropriateness of this approach in Chapters 4 (plant agriculture), 7 (BST), 8 (SCP) and the related extension of those discussions in this chapter.

Barriers to biotechnology transfer

Many of the current genetic engineering breakthroughs clearly hold the potential for poverty alleviation and greater rural labour absorption, but developing countries and their farmers' access to these technologies is limited due to the legal and financial barriers generated by intellectual property rights which create patents on biotechnology products, even though these can be easily developed and adapted to suit developing country conditions. It is quite clear that if the poor cannot afford to buy the biotechnology seeds, the same adverse socioeconomic scenario as was observed for the Green Revolution, will be repeated. It is also clear (from Table A.1) that although the question of what gene to transfer is ultimately a social question, the process is almost entirely being determined by the market at present.

It is not difficult to prevent the entry of socially harmful biotechnologies into the Third World. Tariffs and excise taxes could be imposed to discourage their import and use. The major problem is of enhancing Third World countries' negotiating power to influence their access to the socially beneficial technologies. The current international legal and institutional structures with respect to intellectual property rights certainly tend to erode Third World countries' negotiating power. This unequal relationship is clearly demonstrated by the concrete illustration from Mexico described below.

Mexico, a germplasm-rich country, believes that the world's plant genetic resources constitute a common heritage of mankind. Accordingly, the Mexican patent and certificate of invention law holds that plant varieties and animal species, including new biological procedures cannot be patented. As a consequence, multinationals have free access to Mexico's germplasm without requiring them to provide anything in return (Eastmond *et al.*, 1989). It is little wonder that a multinational from the United States seized this advantage to slip into neighbouring Mexico to test transgenic antisense tomato on Mexican soil (reported in Table A.1). The concept of legal property rights could be expanded to include the Third World's germplasms. Moreover, lack of knowledge on the intricacies of the laws of intellectual property rights and lack of adequate staff greatly restrict Third World countries' bargaining power at formal North-South negotiating fora. A sample of six negotiators from developing and two from the industrialised countries at the Uruguay Round Negotiations under GATT auspices, revealed that none of them had the knowledge of the principles of impact of patent protection on plants (*Biotechnology and Development Monitor*, 1990).[7]

The support to programmes of assistance to Third World countries by the governments of industrialised countries would in the short run also benefit the sophisticated scientists (geneticists) in the industrialised countries in exchange for which the Third World countries could provide the raw material, their most valuable genetic resources.

Further research is needed to empirically examine many more specific Third World country situations to provide policy insights on both national and international measures which could improve the free flow of information and socially beneficial biotechnologies.

Enhancing Third World S&T capacity

Many Third World countries possess the scientific capabilities to deal with the 'second generation' biotechnologies. Based on the review by this volume, further research is needed to identify (a) country-specific measures to strengthen this scientific capacity and, more importantly, (b) channel and direct this limited capability to crops which could contribute to poverty alleviation.

Since multiple cropping was the major determinant of increased labour absorption in the Green Revolution areas, due attention has to be given to the role and contribution of micropropagated potatoes to cropping intensity. Many Third World countries possess this relatively cheaper scientific capacity but it needs to be strengthened. Their work priorities would obviously have to be oriented to focus on labour-intensive food crops of mass consumption. Moreover, such facilities could enhance total employment by blending the sophisticated skills of high-level scientific professionals, primarily women (for the generation of plantlets) with the traditional agricultural labour force. The plight of the underpaid women scientists in the micropropagation laboratories needs to be improved.

Since the high cost of second generation biotechnologies prevents the resource-poor small producers from adopting these technologies, studies need to be conducted on the effect and possibility of larger-scale production of these biotechnologies to reduce the unit cost.

Global impact on nonfarm employment

This volume has attempted an indirect assessment of the positive or negative effects on non-farm employment through the mechanism of backward linkages from agriculture following the introduction of biotechnologies. Due to data limitations the conclusions are obviously tentative and superficial. Therefore, there is a need for quantification of the magnitude of this effect through the generation of hard data by means of concrete empirical case studies.

The cumulative effect of inter-industry repercussions of the resource-saving biotechnologies based on a simulation model is obviously highly speculative. The conclusion that widespread use of biotechnologies could reduce GDP and depress aggregate employment is a serious one. This preliminary simulation exercise based on a theoretical input-output model needs to be more rigorously tested by means of empirical data from a range of developing countries before designing appropriate macroeconomic policies to counter these global effects on the economy.

NOTES

1. Genetic manipulation of many species of potatoes is easier because they carry genes on four sets of chromosomes in each cell as compared to two sets carried by animals and most other plants (*The Economist*, Oct. 1990).
2. Additional incentives created by the favorable economic climate for the promotion of this technology can be seen from the section on SCP to combat malnutrition.
3. Plant used to produce high quality spirits, particularly tequila (high alcohol content drink) which is an important Mexican export.
4. For a description of the SCP technology see Chapter 8.
5. Actually the fear was that BST could transfer from the milk into the blood, particularly in infants, which could produce hormonal and allergic effects. The animal rights lobby has argued that it was an unnecessary and cruel way of squeezing out more milk from the cow to pour into overflowing milklakes in the industrialised countries.
6. Some of these suggestions were made by Hans W. Singer while making a selective review of ILO World Employment Programme Research (Singer, 1991).
7. The developing countries were India, Indonesia, Malaysia, Mexico, the Philippines and Zimbabwe. The government representatives of the two industrialised countries were from the United States and the Netherlands.

TABLE A.1 *Genetic engineering activity and breakthroughs world-wide by agronomic trait, crop, sponsor and potential benefits*

Agronomic trait	Genetic engineering breakthrough	Sponsor	Crops affected	Potential benefits for poverty alleviation
Insect resistance*				
	Immunity from Bollworm and Budworm by incorporating genetic material.[1]	Calgene	Cotton	– Cost saving (pesticides) – Reduces crop losses
	Kills 2 types of caterpillars	Agracetus, Wisconsin		– Nearly a third of worldwide chemical insecticides worth $5 billion are applied to cotton
	Gene inserted from one bacteria into another bacilus thuringienis producing toxin which kills corn borers tunnelling into stalks causing corn ear to fall off.[1] Field trials of the same genetically-engineered micro-organism as was used on corn to protect against rice stem borers proposed at the Ingleside, Maryland Research Farm[2]	Crop Genetics	Corn Rice	– Cost saving – Reduces crop losses
	Genes for natural toxin introduced into tomato kills hornworms when they bite a leaf.[3]	Monsanto	Tomato	(same)

220

Description	Company	Crop	Notes
A gene inserted from a bacteria *bacilus thuringiensis* into another called endophyte and introduced into cracks of corn seed reproduces naturally. When European corn borer bites into it, it cripples the insect's intestinal muscles. They stop eating and die.[4]	Crop Genetics International, Hanover, Maryland	Corn	– Crop damage in the US $400 million despite $50 million expenditure on chemical control – Two-thirds US corn fields affected by it – One season's corn surplus in Kenya turned into a health hazard – Same microbe can colonise 83 other plant varieties – Costs half of chemical sprays[5] – Has potential to spread and benefit poor farmer neighbours
Pea Lectingene from pea inserted into potato interferes with digestive process of pests like colorado beetle or tubermoth. Available as seeds.[6]	Nickerson International Ltd. Norfolk (UK) (Seed Company)	Potato	– Emerging as a major food source in Africa and Asia, particularly India, China, Viet nam and Sri Lanka – Would increase labour use through multiple cropping
Resistance to cotton bollworms, tobacco budworms, beet armyworms.[7]	Monsanto	Cotton Tobacco Beet	– Cost reducing – Crop losses reduced
Gene transferred from Cowpea to tobacco enabled production of protein which disables an enzyme (trypsin) used by tobacco bud worm to digest food. As the insect bites into the transgenic plant, it is unable to digest the food and starves to death.[8]	–	Tobacco	(same)

Agronomic trait	Genetic engineering breakthrough	Sponsor	Crops affected	Potential benefits for poverty alleviation
Microbes for cold tolerance (Cont.)				
	Genetically altered bacteria baculovirus to control insect cabbage looper[33]	Cornell University	Cabbage and a dozen different vegetables	– US authorities authorised field tests
Microbial fungicide				
	Fungal disease attacks roots, afflicts wheat fields around the world, cuts crop yields by 50%. Cultured microbes attack the fungus and can be applied to the wheat seeds.[35]	Monsanto[34] Agricultural Co.	Wheat	– Spread to poor people's soils and fix nitrogen or reduce disease/pests there Crop losses reduced – US Environmental Protection Agency has given permission to Monsanto to field test in Pullman, Washington State
Microbes attack disease				
	Crown gall disease affects stone fruits, nuts and roses caused by bacteria *A. tume faciens* in soil. Genetically engineered bacteria solution (10 billion bacteria packed in one litre) soaked in roots of seedlings. Marketing began by Bio-care Australia in 1989.[36]	Bio-Care *Australia*	Stone fruits, *nuts, roses*	– $150 million losses world-wide – *Could spread to poor neighbour's soils, multiply and protect crop there* – Costs US$1.20 per litre[37] – Bio-Care has already begun marketing this transgenic bacteria known as NoGall in 1989[38]

Transform rice plants within 2 years by genetic engineering to resist especially destructive disease rice tungro virus that stunts growth in rice plants.[12]	University of Nottingham (UK)	Rice	– The virus can devastate whole fields
Gene incorporated into cucumber protects the plant from the cucumber mosaic virus which distorted leaves.[13]	Cornell and New York State Agricultural Experiment Station	Cucumber	– The virus reduces crop yield and quality – Potential benefit for both developed and Third World countries
Russet Burbank potato genetically engineered show resistance to potato viruses X and Y.[14]	Plant Genetics Subsidiary of Calgene, Davis, California	Potato	– Russet Burbank potatoes account for more than 40% of the North American commercially produced potato – Valued at US$2 billion per year[14] – Viruses X and Y reduced crop yields by 10% each year

Herbicide resistance**

Plant grown from single cells carrying a gene for herbicide resistance.[15]	Belgium	Sugar beet	– Displaces labour, especially women – Raises farmer's costs – If poorer farmer's fields suffer increased weed growth without herbicides (pollinating agents move there from the herbicide affected fields) yield there falls[16]

Agronomic trait	Genetic engineering breakthrough	Sponsor	Crops affected	Potential benefits for poverty alleviation
	Imida Zolinone herbicide resistance controlled by a single dominant gene patented in 1988 incorporated into 100 maize lines commercially available in 1992[17]	Pioneer H-Bred International Co. USA (largest seed company)	Maize	(same)
	'Monocot barrier' broken by genetically engineered herbicide resistance to produce transgenic fertile corn.[18]	–	Corn	(same)
	Monsanto scientists have genetically engineered cotton for tolerance to the company's non-selective roundup herbicide[19]	Monsanto	Cotton	(same)
	Gene for herbicide resistance inserted into commercial tobacco[20]	Du Pont	Commercial tobacco	(same)
	Gene transferred for resistance to broad spectrum herbicide Basta[21]	Plant Genetic Systems	Tomato, potato, tobacco	(same)
	Gene inserted into tobacco resistant against Monsanto produced herbicide roundup[21]	Monsanto	Tobacco	(same)

Nitrogen-fixation

		Crops	
Ongoing research to shift 20 genes from nitrogen fixing bacteria into crops.[22]	Sussex, UK	Rice, wheat	– Reduction in average cost of production of poor farmers
For creating symbiosis between cereals and NIF bacteria stimulated nodule-like structures in rice and wheat roots containing rhizobia in 1989.[23]	University of Nottingham (Ted Cocking)		(same)
In 1988 complete sequence DNA in NIF cluster of 20 genes all in a row in *klebsiella pneumoniae*. Worked out how NIF regulated in sufficient detail. Promising source for transfer. This method already used to create new NIF bacteria.[24]	Germany (Alf Phler)		– Reduction in average cost of production of poor farmers

Drought resistance

		Numerous crops	
Since large number of separate traits help withstand drought, it is difficult and complex to isolate and transfer as many as 50 genes. Each of several mechanisms which the plant uses to overcome drought controlled by a set of genes (50) now breaking down traits for drought tolerance into its biochemical and physiological components *viz.* deeper roots, thicker cuticle covering the plant, chemicals in plant that help reduce water loss. Each of these makes a modest contribution but cumulatively could make a major contribution.[24]	–		– Area expansion – Multiple cropping – Cost reduction – Risk averse small farmers feared the higher fluctuations in output for Green Revolution crops compared to lower fluctuations of traditional varieties when the volume and timing of water was not appropriate

Agronomic trait	Genetic engineering breakthrough	Sponsor	Crops affected	Potential benefits for poverty alleviation
	Gene inserted from petunia growing in desert into normal petunia reduced water requirement by 40%.[25]	–	Petunia	(same)
	Genetic engineering will allow the insertion of cactus genes into wheat, corn, or soybeans to produce 'less thirsty' grain crops.[26]	California	Grain	(same)
Processing and canning				
	Modify texture, taste, colour, shape. Roughly 70% tomato crop in the US processed. Commercial processors interested in fleshy and solid parts (95% liquid) of tomato.[27]	H. J. Heinz Co. & Campbell Soup Co.	Tomato	– 1% improvement in the proportion of solid part of tomato would add $77 million to annual value of processed tomato[27]
Bakeries				
	Genetically engineered yeast (gene inserted from another yeast) modified the genes to produce carbon dioxide more quickly, and so make the bread rise faster[28]	Gist Brocade UK	Bread making	– Approved for commercial use
Seedless orange				
	'Gene shears' used to switch off specific genes which lose their functions. Block development of seeds in fruits like citrus.[29]		Fruits	– Could benefit both Third World and industrialised countries – Problem of legal and monetary barriers (associated with patents) to access by LDCs

Pasture crop

Description	Organisation	Crop	Benefits
Gene for sulphur rich amino acid transferred from a pea seed to the leaves of a pasture crop.[30] Sheep grow 30% more wool by feeding on the genetically engineered diet	The Commonwealth Scientific & Industrial Research Organisation (CSIRO), Canberra, Australia	Pasture crops, tropical legumes	– Australian wool production could increase by 5% bringing in an extra Australian $300 million annually – LDCs would benefit through (a) grazing land saving, and (b) higher production

Decorative value

Description	Organisation	Crop	Benefits
Luminosity (luciferane) gene of the fire fly inserted into the tobacco plant that glows in the dark[31]	University of California, San Diego	Tobacco	– Ornamental and commercial value mainly in industrialised countries

Microbes for cold tolerance

Description	Organisation	Crop	Benefits
With the deletion of the gene for ice nucleition protein, ice negative bacteria sprayed on leaves of crops could prevent wild-type ice-positive bacteria from gaining foothold on leaves of crops.[32]	Advanced Genetic Sciences Inc./ University of California, Berkeley	Strawberry Potato	– Spring crops in the US $1 billion in frost damage[33] – Benefits confined to temperate zones – Trials show that treated plants had only 1/3 of the frost damage on unprotected plants and the genetically engineered microbe did not spread beyond 30 metres test area

Agronomic trait	Genetic engineering breakthrough	Sponsor	Crops affected	Potential benefits for poverty alleviation
	Toxic gene extracted from bacterium *bacilus thurriengensis* being inserted into rice within 1991 for protection against rice *stem borers* and *leaf folders*.[9]	IRRI	Rice	– World-wide rice accoutns for one-fifth of humanity's calorie intake; in some Asian countries leading source of calorie intake [10] – Potential yields 6 bushels per hectare as against only 3 bushels currently achieved average [10]
	Genetic engineering is being explored in rice plants to introduce bacterial insecticides that kill insects that feed on rice plants. It has already been applied as a spray for years but considered as expensive and inefficient.[10]	University of Ghent, Belgium	Rice	(same)
Disease resistance				
	Genes for protection against tobacco mosaic virus inserted into chromosome of tomato cell.[11]	Monsanto	Tomato	– Increases yields by 25%

Microbes for nitrogen-fixation

Bacterium Klebsiella oxytoca associated with rice genetically engineered to increase nitrogen content in rice by 30 per cent.[37]	Department of Agriculture, Tokyo University	Rice	– Rice provides half of total calorie intake for 2 billion people in the world; nearly 70 per cent of the protein is provided by rice in the diet of the population of some parts of Asia (Walgate 1990, p. 6)

Quality of crop ripening

Genetically engineered 'antisense' gene into tomato which blocks formation of enzyme involved in the softening of tomato ripening. Field trials ongoing in Mexico reported to be successful.[38]	Calgene funded by Campbell Soup	Tomato	– Given inadequate marketing infrastructure potentially valuable for Third World but patency costs would pose a barrier – Prolongs shelf life (reduced rotting) – Increases total solid content and displays viscosity and consistency – US patency (No. 5,801,540) obtained[38]
Produce tomato ripening characteristics which facilitate canning.[39]	ICI	Tomato	– Of primary interest to developed countries but processors in the Third World could benefit

Microbes destroy insects

Genetic engineering increased the virulence level of a virus 100 fold to control insect *Diatrea Saccharalis*[40]	University de Campinas and PLANALSUCAR Brazil	Sugarcane, soybean, millet, garden vegetables	– Cost of application of the virus is $10 per hectare to prevent crop losses from insect attack to the extent of $100 a hectare[30]
Genetically engineered micro-organism *Bacillus thuringiensis* is two to three times more effective against caterpillars[41]	Repligen Sandoz Research	Various crops	– Reduces costs of farming and replaces chemical insecticides

NOTES

* Cost of producing a new chemical insecticide is 100 times greater than the cost of genetically engineering a crop with a single gene, toxic for a specific pest.
** It is estimated that at least 40 herbicide resistance projects are in progress (Hobbelink, 1989, p. 6).

1. *Financial Times*, 4 April 1989 and *Biotechnology Bulletin*, Vol. 7, No. 10, Nov. 1988.
2. *Chemical and Engineering News*, 6 March 1989, p. 28.
3. *USA Today*, Wednesday 28 September 1988.
4. *The Economist*, 16 April 1988.
5. Sundquist, W.B. *Emerging maize biotechnologies and their potential impact*, Paris, OECD, Technical Paper no. 4, October 1989.
6. *New Scientist*, 8 September 1990.
7. UNIDO, December 1989.
8. *The Daily Telegraph*, 12 November 1987.
9. *Biotechnology and Development Monitor* (The Hague), No. 4, September 1990.
10. *The New York Times*, 6 February 1990.
11. *USA Today*, Wednesday 28 September 1988.
12. *The New York Times*, op. cit.
13. *Chemical and Engineering News*, 28 August 1989, p. 21.
14. *Chemical Week*, 30 August 1989.
15. *New Scientist*, 18 August 1988.
16. Lipton, Michael A., R. Longhurst, *New Seeds and Poor People* (London, Unwin Hyman, 1989), p. 372.
17. Sundquist, op. cit.
18. A dramatic breakthrough has been achieved by which De Kalle Genetics, Biotechmed International (Cambridge Massachusetts) and Monsanto has introduced a gene into corn, the transgenic plant can then be grown producing seeds European Chemical News, 5 February 1990). Although this can be used for numerous manipulations, not surprisingly De Kalle Genetics has started off with application for herbicide resistance while Monsanto has successfully incorporated a gene from the fire fly (Chemical Engineering News, 30 April 1990, p. 26).
19. *Chemical and Engineering News*, 28 November 1988, p. 21.

20. *Chemical and Engineering News*, 2 February 1987, p. 28.
21. *European Chemical News*, 2 February 1987.
22. *New Scientist*, 31 March 1990.
23. *New Scientist* 3 February 1990.
24. *New York Times*, op. cit.
25. *USA Today*, 28 September 1988.
26. *International Herald Tribune*, New York, 17 March 1986.
27. Goodman, D., Sorj, B., Wilkinson, *From Farming to Biotechnology: A Theory of Agroindustrial Development* (Oxford, Blackwell, 1989).
28. Watts, Susan: 'Have we the stomach for engineered food?', in *New Scientist*, 3 Nov. 1990, p. 24.
29. *New Scientist*, 26 May 1988.
30. *New Scientist*, 10 March 1988 and 13 November 1986.
31. *New Scientist*, 6 October 1990.
32. *New Scientist*, 26 May 1988, McGraw Hill's *Biotechnology Newswatch*, 4 May 1987 and *Chemical and Engineering News*, 21 November 1988, p. 26.
33. *The Wall Street Journal*, 29 Mar. 1989.
34. *Biotechnology Bulletin*, Vol. 7, No. 9, Oct. 1988.
35. *International Herald Tribune*, 12 July 1990.
36. *New Scientist*, 4 March 1989 and Greenfield, P.F. 1991, p. 3.
37. *Bio/Technology*, Vol. 8, June 1990.
38. *Chemical Week*, 22 February 1989 and *Scientific American*, May 1990, *Chemistry and Industry*, 19 Sept. 1988 and *Chemical Week*, 13 Sept. 1989.
39. The InterAmerican Development Bank: *Economic and social progress in Latin America*, 1988 Report, Washington D.C., p. 245.
40. *Financial Times*, 4 April 1989.
41. *European Chemical News*, 1 Oct. 1990.

Glossary

Advanced plant biotechnology	Genetic engineering and plant tissue culture techniques as opposed to traditional plant breeding techniques.
Adventitious roots and shoots	Organs developed from any part of a stem except the apex or the axils.
Aflatoxin	A potent toxin produced by certain fungi that commonly grows on damp or poorly dried seeds, grains and nuts.
Amino acid	An organic compound that contains both a basic amino group and an acidic carboxyl group. Basic component of proteins. The chain linkage of amino acids in a particular sequence determines the character of the different proteins.
Antibody	Protein component of the immune system of mammals that is found in the blood and can react in the presence of one or more types of antigen.
Antigen	Usually a protein or carbohydrate which, when introduced into the body of a human being, or of an animal, induces the production of an antibody that reacts specifically with it.
Apical meristem	The rapidly dividing tissue at the apex of a plant responsible for its growth.
Artificial sweetener	Sugar substitutes, either chemically synthesised or not and derived from sources other than the traditional sugarcane, sugarbeet or palms. Sugar mainly refers to the chemical, sucrose.
Assam type tea	A type of tea with very large leaves and soft and large shoots.
Axillary shoots	Shoots that develop from the buds located in the axils of leaves, also called lateral buds.
Bacteria	A minute, unicellular microscopic organism which is found everywhere.
Bacterial wilt	A disease of potatoes caused by bacteria found in the soil which cause the whole plant to wilt. It is hard to eradicate from

an infected soil.

Basic seed — The seed arising from material grown from foundation stock or clones.

Biocatalyst — Enzyme that activates or speeds up biological or industrial processes.

Bio-conversion — The transformation or breakdown of substances from one form to another or to its constituents through the activities of micro-organisms or enzymes.

Biodegradable — Descriptive of substances that are decomposed by the activity of micro-organisms.

Bio-fertilisers — Nitrogen-fixing systems that are used to enhance the nutrient content especially nitrogen of soils and improve crop production.

Bio-fuels — Substances of high energy value like methane and ethanol produced when some micro-organisms degrade organic materials such as sugars, cellulose and the like.

Biogas — The flammable, odourless gas produced by anaerobic micro-organisms actively growing on organic wastes.

biomass — Any organic substance resulting from the photosynthetic conversion of solar energy.

Bio-pesticides — Virus and micro-organisms used to control insect pests and diseases.

Biopolymer — Organic compound with a high molecular weight found in nature including cellulose and the principal components of wood, jute, cotton, etc.

Bioprocess — Any process that uses complete living cells or their components.

Bioreactor — Vessel in which a bioprocess takes place.

Biosynthesis — Production of a chemical compound by synthesis or decomposition caused by living organisms.

Biotechnology — Commercial techniques which use living organisms to make or modify a product, including the characteristics of economically important plants and animals and for developing micro-organisms which act on the environment.

Broilers — Poultry reared for the meat.

Callus	Undifferentiated mass of plant cells that constitutes a first step in the regeneration of plants in the tissue culture technique.
Campesino	The Spanish word for peasant or agricultural worker.
Carcinogenic	Cancer-causing.
Catalyst	Substance that increases the rate of a chemical reaction without being consumed during the process.
Cell	The smallest structural unit of living matter capable of functioning independently; a microscopic mass of protoplasm surrounded by a semipermeable membrane, usually including one or more nuclei and various non-living products, capable alone, or interacting with other cells, of performing all the fundamental functions of life.
Cell culture	The *in vitro* growth of cells isolated from multicellular organisms. These cells are usually of one type.
Cell line	Cells that have the ability to multiply themselves indefinitely *in vitro*.
Cell and protoplast culture	The propagation of cells and cells without cell walls (protoplasts) in the laboratory using a constituted growth medium.
Cellular	That which has to do with the smallest living unit of plants or animals, the cell.
Cellulose	A polymer of six-carbon sugars found in all plant matter; the most abundant biological compound on earth.
Cellulosic agricultural waste	Complex carbohydrate waste material which forms the chief constituent of the cell walls of agricultural products (plants and animals).
Certified seed	Certified seed refers to any material that has been inspected and certified as free from certain levels of diseases and as having been grown under certain specified conditions. Certified seed is usually grown from basic seed or from other certified seed of high quality.
Ceteris paribus	All things being equal.
Chromosome	Unit of the genome made up of linear or circular molecules of DNA with or without the proteins associated therewith.

Characterisation of varieties	The description of the properties of a variety through various tests and observations so that it can be distinguished from other varieties.
China type tea	A type of tea that produces very small, tough shoots which are hard to pluck and which contribute a high percentage of fibre in made tea.
Clonal tea	An improved type of tea that is developed through cross-breeding and selection and propagated by growing tea bushes from leaf cuttings.
Clone	A plant population raised from a single plant part usually through tissue culture and/or vegetative propagation. The individual members of the population have exactly the same genetic characteristics.
Cloning or clonal multiplication	A group of cells or an organism derived asexually from one single cell. Recently cloning in molecular biotechnology refers to the replication of a small DNA molecule or a gene as in cloning vector.
Cotyledons	Seed leaves or food storage parts of a seed.
Crop diversification	The production of different crops in a land area previously planted to a single crop.
Crude protein content	The quantity of nitrogen contained in a compound, or particularly for a protein substance – the factor 6.25 when multiplied by the weight of nitrogen (in grams) derived from a sample containing protein, gives the approximate weight (in grams) of crude protein content in the sample.
Cryopreservation	Preservation and storage of cells, and tissues and organs at very low temperatures by immersion into liquid nitrogen.
Culture (Culture Medium)	System of nutrients required for the artificial cultivation of bacteria and other cells.
Cultureware	Usually refers to disposable plastic or glassware used in tissue culture technology.
Deoxyribonucleic Acid (DNA)	A linear polymer, made up of deoxyribonuceotide repeating units, that is the carrier of genetic information; present in chromosomes and chromosomal material of cell organelles such as mitochondria and chloroplasts, and also present in

some viruses. The genetic material found in all living organisms. Every inherited characteristic has its origin somewhere in the code of each individual's DNA

DNA Sequence
The order of nucleotide bases in the DNA helix; the DNA sequence is essential to the storage of genetic information.

Digestibility
Process by which nutrient materials are rendered soluble and absorbable by action of various juices containing enzymes.

Disease-resistant plants
Plants possessing an inherent capacity to ward off specific disease-causing organisms.

Ejidatarios
Those individuals who have rights in the ejido.

Ejido
The institution through which land has been distributed to peasant communities by the land reform programme since the 1917 Revolution. Land titles are retained by the Federal Government and the members of the community (ejidatarios) are granted lifetime usufruct rights. The ejidos are administered by elected representatives under the supervision of the Ministry of Agrarian Reform.

Electroporation
The use of electric current to 'make' a small pore in the cell membrane to allow DNA or genes to enter the cell.

ELISA serological method
This is a method of testing for the presence of viruses in plant tissue. The acronym ELISA stands for Ensyme-linked Immunosorbent Assay.

Embryo engineering
Inserting foreign genes or cells into embryos so the adult will have the characteristics of the inserted genes or cells.

Embryo Transfer Technology
Micro-injection of isolated DNA into embryo cells and implantation of these embryo cells to which genes have been transferred into surrogate mothers (see 'gene transfer').

Embryogenesis
The formation of embryos from *in vitro* cultured cells or tissues.

Enzyme
Complex protein produced by living cells that is capable of hastening specific biochemical reactions.

Explants
Any part of a plant that is used to initiate a

	culture *in vitro*.
Fatty acids	Organic acids with long carbon chains. Fatty acids are abundant in cell membranes and are widely used as industrial emulsifiers.
Fermentation	A slow decomposition process of organic substances induced by micro-organisms, or by complex nitrogenous organic substances (enzymes) of vegetable or animal origin, usually accompanied by evolution of heat and gas. It is used for the manufacture of such products as alcohols, acids, cheeses, etc.
Fermentor	Equipment for carrying out fermentation.
Foundation stock	This is the first seed material from a newly-developed variety at the final stages of various tests and naming of the variety.
Gamaglobulin	Type of blood protein that plays a predominant role in the immune process.
Gametes	The haploid reproductive cells of either sex.
Gene	The basic unit of heredity; an ordered sequence of nucleotide bases, comprising a segment of DNA. A gene contains the sequence of DNA that encodes one polypeptide chain (via RNA)
Gene expression	The mechanism whereby the genetic directions in any particular cell are decoded and processed into the final functioning product, usually a protein.
Genetic engineering	The manipulation of the contents of plant or animal cells to change their inherited characteristics. (See 'Recombinant DNA Technology').
Genetic manipulation	Refers to various procedures and techniques of changing the heritable make-up of an organism. Although the term may include the traditional methods of mutation and hygridisation, genetic manipulation often connotes the new techniques of protoplast fusion, molecular cloning and genetic engineering, procedures that can be carried out only in the laboratory.
Gene Transfer	The use of genetic or physical manipulation to introduce foreign genes into host

	cells to achieve desired characteristics in progeny.
Generation of Hybridomas	The cell line resulting from the artificial joining of different types of cells (see 'Hybridoma').
Genome	Chromosome complement of an organism or set of hereditary material (genes). It may be composed of one or more chromosomes, depending on the complexity of the organism.
Genotype	Genetic characteristics of an individual or group.
Germplasm	Total variability of genetic material of a particular species.
Gestation	Period from fertilisation of the ovary to birth.
Glycine	Simplest of the amino acids.
Gout	A metabolic disease that is characterised by an increase in the concentration of uric acid in the blood and by its deposition as sodium urate in joints, bones, ligaments and cartilages which might lead to sudden painful swelling of the joints usually of the big toe.
Growth hormone	Hormone produced by the anterior pituitary. Chemically, it is a peptic acid made up of fourteen amino acids. Its principal effect is the stimulation of the growth of the skeleton, but it also plays a role in the metabolism of fats, proteins and carbohydrates.
Growth medium	This is the substance in which a plant or an organism is grown. It may be soil, sand or a carefully constituted and highly specialised substance.
Haploid	Cells containing a single set of chromosomes.
Herbicide	An agent (e.g. a chemical) used to destroy or inhibit plant growth; specifically, a selective weed killer that is not injurious to crop plants.
High Fructose Syrup	A sugar substitute derived from the breakdown of starch into its component, glucose, which in turn is enzymatically converted into its sweeter counterpart, fructose.

Homozygous	Individuals with identical alleles for a specific trait.
Hormone	Product of living cells that circulates in the fluids of both animals and plants and which acts as a chemical messenger inhibiting or stimulating the activities of cells distant from the place in which it is produced.
Hybrid	The offspring of genetically dissimilar parents (e.g. a new variety of plant or animal that results from cross-breeding two different existing varieties, a cell derived from two different cultured cell lines that have fused).
Hybridization	Joining of two complementary strands of RNA or DNA, which gives rise to a double-stranded molecule.
Hybridoma	Product of fusion between myeloma cell (which divides continuously in culture and is 'immortal') and lymphocyte (antibody-producing cell); the resulting cell grows in culture and produces monoclonal antibodies.
Hybrid vigour	The intensified expression of desirable genetic traits that makes a hybrid superior to its parents.
Hydrocarbon	An organic compound consisting of carbon and hydrogen.
Hypocotyl	Embryonic stem portion of an embryo of a seed just below the cotyledon attachment.
In vitro	Descriptive of a living culture in a free cellular system lacking cell membranes.
In vivo	Descriptive of a living culture within a cell or organism.
Indian hybrid tea	A type of tea with moderate sized leaves and with soft and large shoots.
Interferon	Type of protein with a low molecular weight that is important in the immune function and is used because of its anti-viral action.
Isomerase	Enzyme that catalyses the transformation of a compound into its isomer.
Kidney stone formation	Formation of hard aggregate or stone that is found in the kidney and that consist chiefly of inorganic matter.

Laminar Flow Hoods	A hood that will provide a sterile working environment by a sterilised air curtain for tissue culture technology.
Late blight	A fungal disease of potatoes that attacks the whole plant, usually under humid low temperature conditions. It kills the leaves and infects the tubers causing them to rot. Reduction in yield can be 100 per cent under severe conditions.
Laying hens	Poultry reared for egg production.
Lignocellulose	Complex biopolymer that includes the mass of woody plants.
Lines/selections	The materials arising out of a selection process prior to being tested and named as varieties.
Lymphocyte	Type of cell found in the blood, the spleen, the lymph nodes, etc., of higher animals. A subclass of lymphocytes produces and secretes antibodies.
Maintenance or clonal breeding	This refers to a package of techniques used to preserve a variety and its characteristics through clonal or other techniques. The maintenance aspect may also involve developing close substitute lines and/or varieties with similar characteristics.
Mendelian genetics	The theories on inheritance of biological characteristics as espoused by Gregory Mendel in 1865. The principles of inheritence which Mendel discovered have since been shown to apply generally among living things that reproduce sexually.
Meristem	The undifferentiated plant tissue from which new cells arise.
Mutagen	An agent that causes mutation.
Mutant	An organism with one or more DNA mutations, making its genetic function or structure different from that of a corresponding wild-type organism.
Metabolisable energy	The chemical energy changes which occur within a living organism or a part of it which are involved in various life activities.
Metabolite	Product of metabolism.
Microbiology	Branch of biology that deals especially with

the study of the microscopic forms of
life.

Microinjection   The use of a very small diameter, less than
10 micrometre, glass needle, to inject
genes or DNA into cells or embryos.

Micrografting   The *in vitro* grafting of a meristem into a
young stem cultured *in vitro*.

Micro-organism   Any organism that requires a microscope to
be seen.

Micropropagation   Using isolated cells from a plant for propagation with tissue culture technology to
yield identical crops.

Molecular biology   A discipline in the biological sciences that
deals with molecules, such as DNA or
proteins.

Molecular genetics   Study of the nature and biochemistry of
genetic material. It includes genetic engineering technologies.

Monoclonal Antibody   Antibody produced from a single source or
clone of cells that recognises only one
type of antigen.

Mutagenesis   The induction of mutation in the genetic
material of an organism; researchers may
use physical or chemical means to cause
mutations that improve the production of
capabilities of organisms.

Mutation   Any change in the sequence of chemical
composition of the bases contained
in DNA molecules. It alters genetic
material.

Mycorrhiza   Root fungus; an association of fungi with
the roots of trees and other plants that
increases the capacity of a plant to absorb
nutrients from the soil.

Nitrogen-fixation   The catching and transforming of gaseous
nitrogen into a form assimilable by
plants. This process is performed either
by a symbiotic or a nonsymbiotic system.
A symbiotic system consists of plant and
a bacterium living closely together to
fix nitrogen. A nonsymbiotic system
consists of a free living bacterium.

Nitrogen, fixation of   Conversion of atmospheric nitrogen into a
biologically available form by nitrogen
fixing organisms.

Nucleic acids   Long-chain molecules that usually occur

in combination with proteins. The two main types are ribonucleic acid (RNA) and dioxyribonucleic acid (DNA). Each nucleic acid chain is composed of sub-units (monomers) that contain a sugar, a phosphate group, and one of four possible bases. The specific sequences of these sub-units constitute the genetic information of the cell.

Nucellus

Tissue that encloses the ovule first and then the embryo.

Nucleus

A relatively large spherical body inside a cell that contains the chromosomes.

Open pollinated varieties

Corn varieties that maintain their desirable agronomic properties throughout even when grown from seeds produced by natural hybridisation among individuals in the same population.

Organelle

A specialised part of a cell that conducts certain functions. Examples are nuclei, chloroplasts, and mitochrondria, which contain most of the genetic material, conduct photosynthesis, and provide energy, respectively.

Organogenesis

The formation of shoots and roots from *in vitro* cultured tissues.

Organoleptic

Method of systematically testing or assessing the effects of a substance on the human senses, particularly taste or smell.

Patent

A limited property right granted to inventors by government allowing the inventor of a new invention the right to exclude all others from making, using, or selling the invention unless specifically approved by the inventor, for a specified time period in return for full disclosure by the inventor about the invention.

Pathogen

A disease-producing agent, usually restricted to a living agent such as a bacterium or virus.

Pathogenic

Descriptive of an entity that causes disease.

Pathongenicity

A disease producing capacity of a micro-organism, a virus, or other substance.

Phenotypic

Phenotype the morphological and functional characteristics of an organism.

Phloem	The vascular tissues of a plant whose function is to conduct food from the leaves down the stem.
Plant and animal breeding	The production of new plant or animal varieties with desired characteristics through controlled reproductive techniques.
Plant apex	The tip of a growing plant.
Plasmid	Circular segment of DNA that is used as a vector for cloning DNA in the host of cells of bacteria.
Polyclonal tea	An improved type of tea that is developed through cross-breeding and selection and propagated by growing tea bushes from seeds.
Poultry	Domestic birds raised for the production of eggs or meat, such as chickens, turkeys, ducks and geese.
Protein	Complex chemical compounds containing nitrogen, carbon, hydrogen and oxygen, found in every body tissue and living cell and are formed from amino acids.
Protoplast fusion	A laboratory procedure of welding together two cells regardless of origin and in the case of plant and microbial cells whose cell walls have been removed.
Radioimmunoassay	Sensitive method of measuring traces of a hormone (or other biomolecule), based on the ability of a hormone to displace radioactive forms of the hormone from their combination with specific antibodies.
Rancid	The development of unpleasant odours and tastes from fats and oil by the oxidation of the unsaturated fatty acid components.
Recombinant DNA (rDNA)	The hybrid DNA produced by joining pieces of DNA from different organisms together *in vitro*. DNA formed by the conjugation of genes in a new combination.
Recombinant DNA Technology	The use of recombinant DNA for a specific purpose, such as the formation of a product or the study of a gene.
rflp maps	Symbolic representation of the genetic make-up of an organism wherein each

	symbol represents a specific nucleic acid sequence. Rflp refers to restriction fragment length polymorphisms.
Rootstock	Basal portion of the stem to which another piece, the scion, is grafted.
Scion	Shoot or stem which is detached and grafted to a rootstock.
Scaling-up	Traditional stage in the transformation of a process from its experimental scale to an industrial scale.
Selection	The screening of better performing individuals in a plant or animal population for breeding purposes or for eventual naming or adoption as varieties.
Selection trials	Experiments that can distinguish from the rest of the population certain individuals that possess properties of interest to the experimenter.
Single Cell Protein	Microbial protein produced from waste materials.
Sludge	Solid residue remaining after treatment of liquid waste.
Somatic tissue	All the tissues that form the body of a plant or animal with the exception of the reproductive tissue which forms the gametes for sexual reproduction.
Somatostin	Growth hormone.
Somoclonal variation	Heritable differences observed among plants propagated through tissue culture of a single mother plant.
Soyabean meal	A product from soybean for feeding animals.
Species	A taxonomic subdivision of a genus. A group of closely related, morphologically similar individuals which actually or potentially interbreed.
Steriod	Compound composed of a series of four carbon rings joined into a structured unit. Any substance of the group of the lipids that occurs naturally and is essential for life.
Structured lipids and tailored fats	Fatty or oily substances designed to possess specific properties resembling naturally occurring fats and oils.
Substrate	Substance on which an effect is exerted, for example, by an enzyme.

Technology transfer	The movement of technical information and/or materials, used for producing a product or process, from one sector to another; most often refers to flow or information between public and private sectors or between countries.
Tissue culture	Method for the *in vitro* growth and conservation of animal and plant cells, which is applied outside the organisms of which they form part. The process involves the isolation of a piece of tissue or organ of a plant or animal and propagating it aseptically in a container of nutrient medium under controlled environmental conditions.
Tissue or meritestem culture	The growing of new plants from small or microscopic pieces of plant tissue as opposed to growing of plants from seeds or whole plant parts.
Toxin	A substance, produced in some cases by disease-causing micro-organisms, which is toxic to other living organisms.
Toxity	The degree of harmfulness of a substance for an organism, the capacity of a substance to produce injury.
Traditional plant breeding	Plant breeding using natural plant reproductive methods, such as pollination, without the manipulation of plant tissue or cells.
Transgenic	Animals that carry foreign genes. Recently, this term has been used for plant species and microbes as well.
Transgenic plants or animals	Plants or animals which have been genetically manipulated to modify certain characteristics or introduce desired ones.
Uric acid formation	Formation of a white, crystalline substance, present in the urine of all carnivorous animals and the principal end product of nitrogen metabolism in man and other organisms.
Variety	Lines/selections that have been tested, passed and named for certain desired unique characteristics.
Vector	Agent for the transmission of genetic information from one cell or organism to

Glossary

	another. The vectors usually selected are plasmids, although viruses or other bacteria may be used.
Vegetative propagation	Equivalent to tissue culture but in this case whole plant parts instead of microscopic parts are used. Highly specialised growth media are usually not required as is the case with tissue culture.
Viroid	Infectious agents.
Virus	Any of a large group of submicroscopic agents infecting plants, animals, and bacteria and unable to reproduce outside the tissues of the host. A fully formed virus consists of nucleic acid (DNA or RNA) surrounded by a protein or protein and lipid coat.
Ware potatoes	Potatoes grown for consumption.
Wild-type	The most frequently encountered phenotype in natural breeding populations.
Yeast	Micro-organisms (fungus) that reproduce by budding, and are characterised by absence of mycelia and induce alcoholic fermentation of carbohydrates.

Bibliography

Agricultural Policy and Strategy Team: 'Rural poverty and poverty programs in the Philippines', *Agenda for action for the Philippine Rural Sector* (Laguna: University of the Philippines at Los Baños, 1986).

Ahmed, I.: *Technological change and agrarian structure: A study of Bangladesh* (Geneva: ILO, 1981).

Ahmed, I.: 'Technology, production linkages and women's employment in South Asia', *International Labour Review*, vol. 126, no. 1 (January–February 1987).

Ahmed, I.: 'The bio-revolution in agriculture: Key to poverty alleviation in the Third World?', *International Labour Review*, vol. 127, no. 1 (1988).

Ahmed, I.: 'Advanced agricultural biotechnologies: Some empirical findings on their social impact', *International Labour Review*, vol. 128, no. 5 (September–October 1989).

Ahmed, I.: 'Biotechnology and rural labour absorption', in A. Sasson and V. Costarini (eds): *Biotechnologies in perspective: Socio-economic implications for developing countries* (Paris: UNESCO, 1991).

Ahmed, I.: 'Will biotechnology alleviate poverty?', paper presented at the *XXI International Conference of Agricultural Economists*, Tokyo, Japan, 22–29 August 1991.

Akinwumi, J. A. *et al.*: *Economic analysis of the Nigerian poultry industry* (Lagos: Federal Ministry of Agriculture and Water Resources, Federal Livestock Department, 1979).

Alaa el Din, M. N. *et al.*: *The utilization of blue green algae for increasing soil fertilization* (Giza, Egypt: Agricultural Research Centre, 1984).

Aragón, C.: Córdova, V. and Gasem, E.: *Rural employment in the Philippine rice sector*, Agricultural Economics Department, Paper no. 84–28 (Los Baños: IRRI, 1985).

Arroyo, Gonsalo; Waissbluth, Mario: *Dessarrollo biotecnológico en la producción alimentaria de México: Orientaciones de Política* LC/MEX/L. 77 (United Nations, Economic Commission for Latin America and the Carribean (ECLA), March 10 1988).

Arroyo, Gonzalo; Arias, Salvador: *Introducción de la biotecnología en el sistema cañero* (Méexico: Universidad Autónoma Metropolitana, 1987), mimeographed.

Azucar, S. A.: *Estadísticas Azucareras: 1986* (México 1987).

Barker, Randolph: 'Socioeconomic impact of modern biotechnology on world trade and economic development in the developing countries', in *Biotechnology study project papers: Summaries of commissioned papers* (The Hague: ISNAR, 1989).

Bartsch, W. H.: *Employment and technology choice in Asian agriculture* (New York: Praeger, 1977).

Bhatty, I. Z.: *Technological Change and Employment: A Study of Plantations* (Delhi: The MacMillan Company of India Limited, 1978).

Bifani, P.: 'New biotechnologies for food production in developing countries with special reference to Cuba and México', in A. S. Bhalla and Dilmus James (eds): *New Technologies and Development: Experiences in Technology Blending* (Boulder, Colorado, and London: Lynne Reinner Publishers, 1988).

Bifani, P.: 'Biotechnology: Overview and Developments in Latin America', in *Economic and Social Progress in Latin America* (1988 Report) (Washington, D.C.: Inter-American Development Bank, 1988).

Biotech News, vol. 5: 128 (February 1987).

Biotechnology and Development Monitor (Amsterdam), no. 4 (September 1990).

Brannon, J.; Baklanoff, E. N.: *Agrarian Reform and Public Enterprise in Mexico: The Political Economy of Yucatan's Henequen Industry* (Tuscaloosa, Alabama: University of Alabama Press, 1987).

Branton, R.; Blake, J.: 'A lovely clone of coconuts', *New Scientist*, no. 98 (1983).

Brill, Winston J.: 'Genetic engineering applied to agriculture: Opportunities and concerns' in *American Journal of Agricultural Economics*, vol. 68, no. 5 (Dec. 1986).

Bunders, Joske F. G. (ed.): *Biotechnology for small-scale farmers in developing countries: Analysis and assessment procedures* (Amsterdam: VU University Press, 1990).

Buttel, F. H.; B. Kenny and Kloppenburg, J. Jr.: 'The IARCs and the development and application of biotechnologies in development and application of biotechnologies in developing countries', *Biotechnology in International Agricultural Research Proceedings of the Inter-Center Seminar on International Agricultural Research Centers (IARCs) and Biotechnology* (Manilla, Philippines: International Rice Research Institute, 1985).

Buttel, F. H., C. C. Geisler: 'The social impacts of bovine somatotropin: Emerging issues' in Molnar, Joseph J. and H. Kinnucan (eds): *Biotechnology and the new agricultural revolution* (Boulder, Colorado: Westview Press, 1989).

Buttel, F. H.: 'Agricultural research and farm structural change', paper presented at the annual meeting of the Northeast Rural Sociologists (Pennsylvania State University, June 5 1986).

Buttel, F. H.; Barker, R.: 'Emerging Agricultural Technologies, Public Policy, and Implications for Third World Agriculture: The Case of Biotechnology', *American Journal of Agricultural Economics*, vol. 67, no. 5 (December 1985).

Button, J.; Koshba, J.: 'Tissue culture in the citrus industry', in J. Reinert and Y. P. S. Bajaj (eds), *Applied and Fundamental Aspects of Plant Cell Tissue and Organ Culture* (Berlin: Springer-Verlag, 1977).

Chahal, D. S.: 'Bioconversion of biomass into protein-rich animal feed' paper presented at the Forum for Applied Biotechnology (Belgium: State University of Ghent, October 1987).

Chemical News, Wednesday, 3 December 1986.

Chicago Tribune, Wednesday, 22 August 1990.

Clayton, E.: *Agriculture, poverty and freedom in developing countries* (London: The Macmillan Press, 1983).

Corley, R. H. V.: 'Cloning planting material for the oil palm industry', *The Planter*, no. 58 (1983).

Crott, R.: 'The impact of Isoglucose in the International Sugar Market', in S. Jacobsson *et al.* (eds): *The Biotechnological Challenge* (London: Cambridge University Press, 1986).

Department of Agriculture, Government of the Philippines: 'Policies, priorities and medium-term program of action' (Quezon City, Philippines, 1987), unpublished draft.

Dingell, J. R.: 'Biotechnology: What are the problems beyond regulation?', in S. Danem (ed.), *Biotechnology – Implication for Public Policy* (Washington D.C.: The Brookings Institution, 1985).

Döbereinez, J.; Neves, M. C. , Donet, P.; Agostinho D. D. and Duque, F. F.: 'Rhizobium strain effects on nitrogen transport and distribution in soyabeans', *Journal of Experimental Botany*, vol. 36, no. 169 (August 1985).

Dohlman, C. J.; Ross-Larson, B.; Westphal, L. E.: 'Managing technology development: Lessons from the newly industrializing countries', *World Development*, vol. 15, no. 6 (1987).

Doyle, J.: *Altered harvest: Agriculture, genetics and the fate of the world's food supply* (New York: Viking, 1985).

Dupuis, M.: 'No limits to growth', *California Farmer*, vol. 263, no. 3 (7 September 1985).

Eastmond, A.; Gonzalez, R. L.; Saldana H. L. and Robert, M. L.: *Towards the application and commercialisation of plant biotechnology in Mexico* (México: Universidad Autónoma de Yucatán, October 1989), unpublished draft.

ECLA: *Desarrollo biotecnológico en la producción alimentaria de México: Orientaciones de política*, Report prepared by G. Arroyo and M. Waissbluth (México D. F., 1988), mimeographed.

Edozien, J. C. *et al.*; 'Effects of yeast feeding on uric acid metabolism in young men', in *Nature*, vol. 228 (1970).

Egbogah, E. O.; Aikhionbare, D. O.: 'Possible new oil potentials of the Niger Delta', *The Oil and Gas Journal* (14 April 1980).

Egbogah, E. O.: 'Nigeria looks to gas for a solution to its problems', *World Oil* and *Opec Bulletin*, vol. IX, 48 (1978).

Ewwll, P. T.: *Intensification of peasant agriculture in Yucatán*, Cornell/International Agricultural Economics Study, A.E. Research 84–4 (Ithaca, New York: Cornell University).

FAO: *Soils Bulletin*, Nos. 40 and 41 (Rome, 1977 and 1978).

Farrington, John (ed.): *Agricultural biotechnology: Prospects for the Third World* (London: Overseas Development Institute, 1989).

Feder, G.; Just, R. E.; Zilberman, D.: 'Adoption of agricultural innovations in developing countries: A survey', *Economic Development and Cultural Change*, vol. 33 (1985).

Fishel, Walter L. 'The economics of agricultural biotechnology' in *American Journal of Agricultural Economics*, vol. 69, no. 2 (1987).

Fowler Cary; Eva Lachkovics; Pat Mooney and Hope Stand: *The laws of life: Another development and the new biotechnologies* (Uppsala, Sweden: The Dag Hammerskjold Centre, 1988).

Fujii, J. A. A.; D. T. Slade; K. Redenbaugh and K. A. Walker: 'Artificial seeds for plant propogation', in *TIBTECHS*.

Genetic Engineering News, February 1989.

Gilliland, Martha: 'A study of nitrogen fixing biotechnologies for corn in Mexico' in *Environment*, April 1988.

Glatz, B. E.; E. G. Hammond; K. H. Hsu *et al.*: 'Production and modification of fats and oils by yeast fermentation' in Ratledge, C., P. Dawson and J. Rattray (eds): *Biotechnology for the oils and fats industry* (American Oil Chemists' Society, 1984).

Gobierno Constitucional de los Estados Unidos Mexicanos: *Programa Nacional de Desarrollo Tecnológico y Científico* (México D. F., 1984).

Goldstein, Daniel J.: 'Ethical and political problems in Third World biotechnology', *Journal of Agricultural Ethics*, vol. 2 (1990).

Gonzaga, V., L.: *Crisis and challenge in Negros Occidental: A study of the sugarcane workers differential perceptions and responses* (La Salle Bacolod City, Philippines: Social Research Center, December 1987).

Goodman, David; Bernardo Sorj and John Wilkinson: *From farming to biotechnology: A theory of agro-industrial development* (Oxford, UK: Basil Blackwell, 1987).

Greenfield, P. F.: 'Developing public sector–private enterprise links in biotechnology: Exercise in S.E. Asia and Australia', in A. Sasson and V. Costarini, op. cit.

Hacking, A. J.: *Economic Aspects of Biotechnology*, Cambridge Studies (Cambridge: Cambridge University Press, 1986).

Hamdan, I. Y.: 'Production of single cell protein from hydrocarbon feedstocks', in *Proceedings of the International Symposium on Single Cell Protein from Hydrocarbons for animal feed* (Algiers, 17–19 October 1983).

Hameed, N. D. A. *et al.*: *Rice revolution in Sri Lanka* (Geneva: United Nations Research Institute for Social Development, 1977).

Hamer, G.: 'Technical aspects of Single Cell Protein production from natural gas (methane), in F. G. Overmine (ed.): *Microbial bioconversion systems for food and fodder production and waste management* (Kuwait Institute for Scientific Research, 1977).

Hayami, Y., and V. W. Ruttan: *Agricultural development: An international perspective* (Baltimore and London: Johns Hopkins University Press, 1985).

Higgins, T. J.; D. J. Best and J. Jones: *Biotechnology Principles and Applications* (Oxford: Blackwell Scientific Publications, 1985).

Hobbelink, H.: 'New hope on false promise: Biotechnology and the Third World agriculture', in *International Coalition for Development Action* (Brussels, March 1987).

Hobbelink, H. 'Agricultural biotechnology and the Third World: International context, impact and policy options' in *Development related research: The role of the Netherlands* (Groningen: University of Groningen, 30 March 1989).

Hobbelink, H.: *Biotechnology and the future of world agriculture* (London: Zed Books, 1991)

Homann, J. *et al.*: *Ware potato storage at the small-scale-farm level: Final report to the Government of Kenya* (Nairobi: Ministry of Agriculture, 1979).

Idachaba, F. S.: 'Beyond self-sufficiency in food production: Programme for storage, processing and marketing, 1986–1990', Paper presented at the Technical co-ordination meeting held at Owerri, 26–27 March 1983 (Lagos: Federal Ministry of Agriculture).

Illo, J. F. I. and Veneración, C. C.: 'Women and men in rainfed farming systems', paper presented at the Women in Rice Farming Systems Workshop, International Rice Research Institute, Manila, Philippines, 1988.

ILO: *Application of modern agricultural technology*, Report, VI, International Labour Conference, 78th Session (1991) (Geneva, 1991).

ILO: *Rural Employment promotion*, Report VII, International Labour Conference, 75th Session (1988) (Geneva 1988).

ILO: *The challenge of employment: Rural labour, poverty and the ILO* (Geneva: ILO, 1988)

ILO:*Structural adjustment and its socioeconomic effects in rural areas*, Advisory Committee on Rural Development, Eleventh Session (Geneva: ILO, 1990).

InterAmerican Development Bank: *Economic and social progress in Latin America*, 1988 Report (Washington, D.C., 1988).

International Food Policy Research Institute: *Accelerating food production in Sub-Saharan Africa* (Washington D.C.: IFPRI, 1986).

International Herald Tribune, 25–26 August 1990.

International Monetary Fund: 'Sustained Price Weakness Forecast for non-Fuel Primary Commodities', *IMF Survey* vol. 16 (13 July 1987) no. 1 (Washington D.C.: International Monetary Fund).

International Service for National Agricultural Research: *Kenya agricultural research strategy and plan – vol. 2: Priorities and programmes* (The Hague: ISNAR, 1985).

International Trade Centre UNCTAD/GATT: *Cocoa: A Trader's guide* (Geneva, 1987).

Ishaque, M.; Chahal, D. S.: 'Crop residues', in D. S. Chahal (ed.), *Food, feed and fuel from biomass* (New Delhi: Oxford & IBH Pub. Co., 1987).

Jain, H. K.: 'Agriculture of tomorrow – Greater productivity, efficiency, and diversity', in *Biotechnology in international research* (Los Baños, Philippines: International Rice Research Institute, 1985).

James, Clive and G. Persley: 'Private/public sector collaboration in biotechnology for developing countries in World Bank/ISNAR/AIDAB/ACIAR *Biotechnology study projects papers: summaries of commissioned papers* (The Hague: ISNAR, 1989).

Johnston, B. F. and P. Kilby: *Agriculture and structural transformation: Economic strategies in late-developing countries* (London: Oxford University Press, 1975).

Jones, K. A.: *Gradient of biotechnologies*, International Centre for Research in the Semi–Arid Tropics (ICRISAT) (Hyderabad, India: ICRISAT, 1987), mimeographed.

Jones, L. H.: 'The use of clonal oil palms in developing countries', in Von Hemert, P. A. Lelieveld, H. L. M. and La Rivera, J. W. M. (eds): *Biotechnology in developing countries* (Delft University Press, 1987).

Junne, G.: 'The impact of biotechnology on international trade', in A. Sasson and V. Constarini, op. cit.

Kalter, Robert J. *et al.*: *Biotechnology and the dairy industry: Production costs and commercial potential of bovine growth hormone*, Department of Agricultural Economics, AE Research 84–22 (Ithaca, New York: Cornell University, 1984).

Kalter, R. J.: 'Statement of Robert J. Kalter, Professor and Chairman of the Department of Agricultural Economics, Cornell University, before the House Subcommittee on Investigation and Oversight of the Science and Technology Committee', United States Congress, April 1985.

Kalter, R. J. and W. Magrath: 'Biotechnology: Economic challenges and opportunities for agriculture, in D. G. Butcher (ed.) *New York Agriculture 2000* (New York: Governor's Office, 1985).

Kalter, Robert J. and Lauren W. Tauer: 'Potential economic impacts of agricultural biotechnology' in the *American Journal of Agricultural Economics* vol. 69, no. 2 (1987).

Kenmore, Z. F.: 'Women, households and farmings systems in an upland area: Misamis Oriental, Philippines', paper presented at the Women in

Rice Farming Systems Workshop, International Rice Research Institute Manila, Philippines, 1988.

Kenya, Republic of: Sessional Paper no. 1 of 1986 on *Economic management for renewed growth* (Nairobi: Government Printer, 1986a).

Kenya, Republic of: *Economic Survey 1986* (Nairobi: Government Printer, 1986b).

Kenya Tea Growers' Association: *Kenya's tea estates*, Information pamphlet (Nairobi: undated).

Keya, S. O.; Freire, J., and DaSilva, E. J.: 'MIRCENs: Catalytic tools in agricultural training and development', in *Impact of Science on Society* (Paris: UNESCO, no. 142).

King, A.: 'Global consequences of new technnologies', in *ATAS Bulletin*, Issue no. 4 (October 1987).

Kinnucan, H.; J. J. Molnar and U. Hatch: 'Theories of technical change in agriculture with implications for biotechnologies' in Molnar, Joseph J. and H. Kinnucan (eds): *Biotechnology and the new agricultural revolution* (Boulder, Colorado: Westview Press, 1989).

Kline, S. J.; Rosenberg, N.: 'An overview of innovation', in R. Landau and N. Rosenberg (eds): *The Positive Sum Strategy: Harnessing Technology for Economic Growth* (Washington D.C.: National Academy Press, 1986).

Kristiansen and Bulock: 'Developments in industrial fungal biotechnology', in *Fungal Biotechnology* (1980).

Kumar, N.: 'Biotechnology revolution and the Third World: An overview' in *Biotechnology Revolution and the Third World* (New Delhi: Research and Information System for the Non-Aligned and the other Developing Countries, 1988).

Lacy, William B. and Lawrence Busch: 'The changing division of labour between the university and industry: The case of agricultural biotechnology', in Joseph J. Molnar and Henry Kinnucan (eds): *Biotechnology and the new agricultural revolution* (Boulder, Colorado: Westview Press, 1989).

Lary, H. B.: *Imports of manufactures from less developed countries* (New York: National Bureau of Economic Research, 1968).

Laskin, I. A.: 'Single-cell Protein', *Annual Report, Fermentation Processes 1* (1977).

Leontief, W. W.: *The Structure of the American Economy, 1919–1939* (London and New York: Oxford University Press, 1949).

Leontief, W. W.: *Input-output Economics* (London and New York: Oxford University Press, 1966).

Levin, R.; Gaba, V.; Tal, B.; Hirsch, S.; Denola, D.: 'Automated plant tissue culture for mass propogation', *Bio/Technology*, no. 6 (1988).

Lipton, Michael A.; R. Longhurst: *New Seeds and Poor People* (London: Unwin Hyman, 1989).

Litchfield, John H.: 'Single cell proteins', in P. H. Abelson (ed.): *Biotechnology and Biological Frontiers* (Washington, D.C.: American Association of the Advancement of the Sciences, 1984).

López Huebe, R.; García de Fuentes, A.: *Manual de información básica de la región henequenera de Yucatán* (A. C., Mérida, México: Centro de Investigación Científica de Yucatán, 1984).

Lu, Yao-chi: 'Impacts of technology and structural change on agricultural economy, rural communities and the environment', in the *American Journal of Agricultural Economics*, vol. 67, no. 5 (December 1985).

Madrigal Lugo, R.; Bailón, R.: 'Aplicaciones de la biotecnología en cultivos agroindustriales: El caso de café', in *La Agroindustria en México*, 1 Seminario Nacional sobre la Agricultura en México (México: Universidad Autónoma de Chapingo, 1987).

Magrath, W. B.; Tauer, L. W.: 'The economic impact of BGH on the New York state dairy sector: Comparative static results', *Northeastern Journal of Agricultural and Resource Economics*, vol. 15, no. 1 (April 1986).

Majisu, B. N.: 'Role of the Potato in Kenya and Government involvement in regional seed potato production courses', in S. Ng'ang'a and R. Shideler (eds): *Potato seed production for Tropical Africa* (Lima: International Potato Centre, 1982).

Malawi Government, Ministry of Agriculture and Natural Resources: *Smallholder Tea Growers in Mulanje: Agro-Economic Survey Report no. 19* (Lilongwe: Ministry of Agriculture and Natural Resources, 1976).

Malawi Government, Office of the President and Cabinet: *Economic Report 1986* (Lilongwe: Department of Economic Planning and Development, 1986).

Manandhar, A.; S. Rajbhandari; P. Joshi and S. B. Rajbhandari: 'Micropropagation of potato cultivars and their field performance' in *Proceedings of national conference on science and technology* (Kathmandu: Royal Nepal Academy of Science and Technology, 1988).

Mani, Sunil: 'Biotechnology research in India: Implications for Indian public sector enterprises' in *Economic and Political Weekly* (Bombay) 25 August 1990.

Mariscal, E. B.; Viniegra-González, G.: 'Cost analysis of yeast protein and RNA production by aerobic fermentation of cane molasses, *Biotechnology and bioengineering symposium*, no. 7 (1977).

Martel, Armando: 'Possible impacts of biotechnology on Venezuela's agroindustry', in A. Sasson and V. Costarini, op. cit.

Martínez-Morales, G.; Villareal Ruíz, C.; Madrigal Lugo, R.: 'Comportamiento en campo de cafetos propagandos *in vitro*', Resumenes, XII Congreso de Fitogenética, 18–22 July (Chapingo, México, 1988).

Maskus, K. E. 'Large costs and small benefits of the American sugar programme' in *The World Economy*, vol. 12, no. 1 (1989).

Mayer, L. V. : 'Biotechnology and U.S. agricultural trade', in *AgBiotech 88*, Proceedings of the conference held in Washington D.C., 26–28

January 1988.

McGraw Hill Biotechnology, no. 9, April 3 (1989).

Mellor, J. W., 'Global food security strategies: Evolution and role' in *World Development*, vol. 16, no. 9 (September 1988).

Mendoza, E. M. T.: 'An assessment of the contribution of national plant breeding programs to nutrition in the Philippines', paper presented at the Nutrition and Welfare Workshop', Working Paper no. 85–1733 (University of the Philippines at Los Baños, October 1985).

Mix, Lewellyn S.: 'How will biotechnology impact upon the dairy industry: What adjustments are ahead?' paper presented at the Ohio Dairy Farmers Federation (Columbus, 13 January 1986).

Moll, H. A. J.: The economics of oil palm. Pudoc Wageningen 1987. The Philippines country profile, The Economist Intelligence Unit (U.K. 1987–88).

Molnar Joseph H. and Henry Kinnucan: 'Introduction: The biotechnology revolution', in Molnar, Joseph J. and H. Kinnucan (eds): *Biotechnology and the new agricultural revolution* (Boulder, Colorado: Westview Press, 1989).

Moscardi, F.: *Utilizaçao de Baculovirus Anticarsia para o Controle da Lagarta-De-Soja, Anticarsia Gemmataus*, Centro Nacional de Pesquisa da Soja/EMBRAPA, Comunicado Técnico, no. 3 (1983).

Mushita, A. T. 'The impact of biotechnology in developing countries' in *Development* (Rome: Society for International Development, 1989 2/3).

Navarro, L.; Juárez, J.: 'Elimination of citrus pathogens in propagative budsooe, II. *In vitro* propagation', in *Proceedings of the International Society of Citriculture*, no. 3 (1977).

Navarro, L.; Ortiz, J. M.; Juárez, J.: 'Aberrant citrus plants obtained by somatic embryogenesis of nucelli cultured *in vitro*', in *Horticultural Science*, vol. 20 (1985)

Nelson, N. D. *et al.*: *Area handbook of Nigeria* (Washington D.C.: U.S. Printing Office, 1972).

New Scientist, 24 March 1988.

New Scientist, 14 July 1988.

New Scientist, 5 January 1991.

Njoroge, I. N.: 'History of research activities at the potato research station, Tigoni', in S. Ng'ang'a and F. Shideler (eds): *Potato seed production for Tropical Africa* (Lima: International Potato Centre, 1982).

Nolasco, M.: *Café y Sociedad en México* (México: D.F. Centro de Ecodesarrollo, 1985).

Norrel, J. R. *et al.*: 'Phillips/Peresta – Single cell protein process', in *Proceedings of the International Symposium on Single Cell Protein from Hydrocarbons for animal feed* (Algiers, 17–19 October 1983).

Norris, J. R.: 'Single cell protein production', in J. R. Norris and M. H. Richmond (eds): *Essays in Applied Microbiology* (New York: John Wiley, 1981).

Nyirenda, H. E.: *Improvement of Tea Production in Malawi by Clonal Selection* (Mulanje, Tea Reasearch Foundation), Mimeographed.

Olubajo, F. O.: 'Production of meat from ruminant animals', *Nigerian Journal of Animals Production*, vol. 3, no. 1 (1976).

Osakwe, E. N. C.: *Practical energy options for Nigeria*, a paper presented at the workshop on Indigenous Technology Development organised by the Anambra State Ministry of Science and Technology, Enugu, 29–31 October 1981.

OTA (Office of Technology Assessment, US Congress): *Technology, public policy and the changing structure of American agriculture* (Washington D.C.: OTA, 1986).

Otero, Gerardo: 'Biotechnology: Employment and environmental challenges for the Third World', in in A. Sasson and V. Costarini op. cit.

Otero, Gerardo: 'The coming revolution of biotechnology: a critique of Buttel', In *Sociological Forum*, 1990.

Padolina, W.: *Brief history of BIOTECH's operations 1980–1984* (Laguna: University of the Philippines, National Institute of Biotechnology and Applied Microbiology, 1985).

Palmer, I.: *The new rice in Asia: Conclusions from four country studies* (Geneva: United Nations Research Institute for Social Development, 1976).

Panchamukhi, V. R. and N. Kumar: 'Impact on commodity exports' in *Biotechnology Revolution and the Third World* (New Delhi: Research and Information System for the Non-Aligned and other Developing Countries, 1988).

Panchamukhi, V. R. (ed.): *Biotechnology Revolution and the Third World* (New Delhi: RIS, 1988).

PCARRD: 'Philippine recommendations for fertiliser usage', *PCARRD Technical Bulletin no. 52* (Los Baños, Laguna, 1983).

Persley, Gabrielle (ed.): *Agricultural biotechnology opportunities for international development: Synthesis report*, World Bank/ISNAR/AIDAB/ACIAR (Washington, D.C.: The World Bank, May 1989).

Persley, Gabrielle: *Agricultural biotechnology: Opportunities for international development* (The Hague: International Service for National Agricultural Research, 1989).

Phillips, Michael J.: 'Microeconomic impacts of emerging technologies', in the *American Journal of Agricultural Economics*, vol. 67, no. 5 (December 1985).

Protein Advisory Group of the United Nations System: Guideline 15, *Nutritional and safety aspects of protein sources for animal feeding, food and nutrition*, Bulletin United Nations University, 4 (3) (7 June 1974).

Quintero, R. (ed.): *Prospectiva de la Biotecnología en México* (México: Fundación Javier Barros Sierra and CONACyT, 1985).

Raines, L. J.: 'The mouse that roared', *Issues in Science and Technology*, no. 4, vol. 4 (1988).

Rajbhandari, S. B.: 'Plant tissue culture method and its potential' in *Proceedings of national conference on science and technology* (Kathmandu: Royal Nepal Academy of Science and Technology, 1980).

Rao, N. S. Subba: 'Nitrogen fixation: can it be exploited?' in *Agri-energy Round Table* (Washington D.C., 1984).

Ratledge Colin, Peter Dawson and James Rattray: *Biotechnology for the oils and fats industry* (American Oil Chemists' Society, 1984).

Rawat, K. J.; Neutyal, J. C.: 'Availability of forest biomass' in D. S. Chahal (ed.), *Food, feed and fuel from biomass* (New Delhi: Oxford & IBH Pub. Co., 1987).

Reyes, J. G.; Neme, J.: 'Vive la industria lechera la peor crisis de su historia: Haro M.', *Excelsior* (México City, 24 May 1988).

Rivera, F. T. and Viloria, V. C.: 'Participation of rural women and children in handwatering agriculture for crop diversification: The case of Luzon, Philippines', *Proceedings of the Third Annual Scientific Meeting, Federation of Crop Science Societies of the Philippines* (Laguna, Philippines: U.P. Los Baños, College, 1987).

Roca, W.: 'Biotechnología: Opportunidades para la investigación agricola en América Latina', in *Fortalecimiento de la Investigación Agricola en Améric Latina y el Caribe* (México, CIMMYT, 1985).

Roca W.: 'In vitro clonal propagation to eliminate crop diseases', in *Biotechnology in International Agricultural Research* (Los Baños: IRRI, 1984).

Rosenberg, N.: *Inside the black box: Technology and economics* (Cambridge: Cambridge University Press, 1982).

Ruivenkamp Guido: The impact of biotechnology on international development: Competition between sugar and sweeteners, in *Viertel jahresberichte des Forschungsinstituts der Friedrich-Ebert-Stiftung*, no. 103 (March 1986), Special Issue on *New technologies and Third World Development*.

Ruivenkamp Guido and Henk Hobbelink: Biotechnologie en de Derde Wereld: De ontmaskering van een nieuwe belofte (Biotechnilogy and The Third World: Unmasking a new promise) in : Derde Wereld 2/1986.

Rural Advancement Fund International (RAFI): *Biotechnology and vegetable oils: Focus on oil palm*, RAFI Communique (Brandon, Manitoba, June 1988).

Sangalang, J.: 'The coconut replanting program', paper presented at the Rice Economy and Agricultural Trade Policy Workshop' (Centre for Policy and Development Studies, University of the Philippines, Los Baños, 3–4 May 1985).

Sasson, A.: *Biotechnologies and Development* (Paris: UNESCO, 1988).

Sasson, A. and V. Costarini (eds), *Biotechnologies in perspective: Socioeconomic implications for developing countries* (Paris: UNESCO, 1991).

Sasson, A.: 'A challenge for the developing World', *UNESCO Courrier*, March 1987.

Sau, R.: 'The Green Revolution and industrial growth in India', *Economic and Political Weekly*, vol. XXIII, no. 16 (16 April 1988).

Schneider, Keith: '5 big chains bar milk produced with aid of BST', in the *New York Times*, 24 August 1989.

Schwarts, R. D. and Leathen, W. W.: Petroleum microbiology, in Miller and Litalcy (ed.) *Industrial Microbiology* (New York: McGraw Hill Book Company, 1978).

Science, 'Gene-watchers' feast served up in Toronto', *News and Comment*, vol. 242, October 1988a.

Science 'Preparing the ground for biotech tests' News and Comment, vol. 242, October 1988b.

Scott, R. H.: 'Research and development in relation to SCP production in the laboratory and industrial scale', in I. Y. Hamdan (ed.), *Proceedings of the International Symposium on Single Cell Protein from Hydrocarbons for animal feed*, Algiers, 17–19 October 1983 (Kuwait: Kuwait Institute for Scientific Research, 1983).

Selignam, R.: 'Single cell protein of new source of food from hydrocarbon fermentation', *Chemical Industry*, no. 13 (1976).

Senez, J.: *The economical aspects of SCP production from petroleum derivatives* (Marseilles, France, 1985).

Senez, J. C.: 'Potentials of single cell protein', in *Proceedings of the 5th International Conference on Global Impacts of Applied Microbiology* (Paris: UNESCO, 21–26 November 1977).

Singer, Hans W.: *Research at the World Employment Programme: Future priorities and selective assessment* (Geneva: ILO, 1991).

Singh, Ajit and H. Tabatabai: 'Facing the crisis: Third World agriculture in the 1980s' in the *International Labour Review*, vol. 129, no. 4 (1990).

Smallholder Tea Authority: *Smallholder Tea Authority*, Annual Report 1984–85 (Thyobo, Malawi, STA 1986).

Society of International Development (SID): Special Issue of Development on Biotechnology, Rome, July 1988.

Sondahl M. R.; W. R. Sharp and D. A. Evans: 'Applications for agriculture. The potential for the Third World', in: *ATAS Bulletin 1. Tissue Culture Technology and Development* (New York: United Nations Centre for Science and Technology for Development, 1984).

Sondahl, M. R.; Nakamura, T.; Medina-Filho, H. P.; Carvalho, L. C.; Fazuoli, L. C.; Costa, W. M.: 'Coffee', in P. V. Ammirato, D. A. Evans, W. R. Sharp, and Y. Yamada (eds): *Handbook of Plant Cell Culture*, vol. 3, *Crop Species* (New York: MacMillan Publishing Co., 1984).

Spiegel-Roy, P.; Vardi, A.: 'Citrus', in P. V. Ammirato, D. A. Evans, W. R. Sharp and Y. Yamada (eds): *Handbook of Plant Cell Culture, vol. 3; Crop Species* (New York: Macmillan Publishing Co., 1984).

Steinkraus, K. H. et al: 'Studies on Tempeh an Indonesian fermented soybean food', Food Research 26, 1960. Quoted in *Microbial Processes: Promis-*

ing Technologies for Developing Countries (Washington, D.C.: National Academy of Sciences, 1979).

Sun, M.: 'Will growth hormone swell milk surplus?', in *Science*, vol. 233 (1986).

Svarstad, H.: *Biotechnology and the international division of labour* (Oslo: University of Oslo, Institute of Sociology, 1988).

Swaminathan, M. S. 'Biotechnology research and Third World agriculture', *Science*, no. 218 (1982).

Swaminathan, M. S.: 'Perspective of biotechnology: Research from the point of view of Third World countries', in *Priorities of Biotechnology Research for International Development*, Proceedings of a workshop (Washington, D.C.: Bostid, National Academy Press, 1982).

Tangley, L.: 'Biotechnology on the farm', *BioScience*, vol. 36, no. 9 (1986).

Tangley, L.: 'Beyond the Green Revolution', *BioScience*, vol. 37, no. 3 (1987).

Tanis, P.: *Biotechnologie in de olien en vettensektor en het effekt op de filippijnse kokosboeren* [Biotechnology in the oil and fats industry and its effect on Philippine coconut growers] (Universiteit van Wageningen, December 1987).

Tea Research Foundation: *Annual Report of the Director to the Annual General Meeting of the Tea Association of Malawi held on 25th April 1987* (Mulanje: Tea Research Foundation 1987).

Terry, E. R.; Akoroda, M. O; Arene, O. B. (eds): *Tropical root crops* (Ottawa: International Development Research Centre, 1987).

The Economist: 'Seed firms: Fruit machines', London, 15–21 August 1987.

The Economist: 'Let the sky rain potatoes', London, 13 October 1990.

The Economist: 'The slow march of technology', London, 13 January 1990.

de la Torre, Mayra: 'Producción de proteinas a partir de residuos lignocelulósicos, *Ciencia y Desarrollo*, no. 37, Año VII (México, 1981).

de la Torre, Mayra y Flores Cotera: *Technical and Economic Evaluation of Different Substrate Alternatives for SCP Production in Mexico*, Symposium on Microbial Biomass Protein, University of Waterloo, Canada, 1985.

Tudge, C.: *Food Crops for the Future: The Development of Plant Resources* (Oxford: Blackwell, 1988).

UNCSTD: 'Tissue culture technology and development', *ATAS Bulletin* (New York: UNCSTD, November 1984).

UNDP: *Programme Advisory Note: Plant biotechnology including tissue culture and cell culture* (New York: UNDP, July 1989).

Ufficio del Ministro per il coordinamento della ricerca scientifica e tecnologica, *Comitato Nazionale per le Biotecnologie*, 1st report, Rome, 1986.

UNIDO: *Genetic engineering and biotechnology monitor* (Vienna), Issue no. 26 (December 1989).

Unnevehr, L. J., Stanford, M. L.: 'Technology and the demand for women's labour in Asian rice farming', in *Women in rice farming* (Aldershot: Gower, 1985).

United States Office of Technology Assessment: *Impacts of applied genetics* (Washington D.C.: United States Government Printing Office, 1981).

United States Office of Technology Assessment: *Technology, public policy and the changing structure of American agriculture* (Washington D.C.: United States Government Printing Office, 1986).

United States Office of Technology assessment: *Commercial biotechnology: An international assessment* (Washington D.C.: Congress of the United States, 1984).

Uyen, N. V. and P. V. Zaag: 'Potato production using tissue culture in Vietnam: The status after four years', in *American Potato Journal*, vol. 62 (1985).

Vaughan, C.: 'The blooming of microbial ecology', *BioScience*, vol. 38 (1988).

Vergopoulos K.: *L'impact des nouvelles technologies sur les industries alimetaires europeennes*. FAST paper no. 26 (Brussels: Commission of the European Communities, Directorate–General for Science, Research and Development, December 1986).

Wald, S.: 'The biotechnological revolution' in *OECD Observer* (Paris), no. 156 (February/March 1989).

Walgate, Robert: *Miracle or menace? Biotechnology and the Third World* (London: Panos Publications, 1990).

Watanabe, S.: 'Employment and income implications of the bio-revolution: a speculative note', *International Labour Review*, vol. 3, no. 124 (1985).

Waterhouse, Price: *Survey of the tea industry in Malawi* (Blantyre: Tea Association of Malawi, 1985).

Webber, I. L.: 'Social organization and change in modern Yucatán', in E. H. Mosley and E. D. Terry (eds): *Yucatán a World Apart* (Alabama: University of Alabama Press, 1980).

World Bank: *World Development Report 1982* (Washington D.C., 1982).

World Bank: *World Development Report 1990* (Oxford: Oxford University Press, 1990).

Yoxen, E. & Vittorio de Martino: *Biotechnology in future society: Scenarios and options for Europe* (Aldershot: Dartmouth, 1989).

Ma Yuanliang: *Modern plant biotechnology and structure of rural employment in China* (Geneva: ILO, 1989), draft.

Zamora, Alfinetta B. and R. C. Barba: 'Status of tissue culture activities and the prospects of their commercialisation in the Philippines', in *Australian Journal of Biotechnology*, vol. 4, no. 1 (January 1990).

Zylberstajn, F. D.; Maria Aparecida Sanches; D. A.Fonseca: A. J.Braga do Carmo; L. Moricochi and P. Campos Torres de Carvalho: 'Estudo economico do uso de processos biotecnológicos na agricultura', [Economic study of the use of biotechological processes in agriculture] in *Biotecnologia e desenvolvimento nacional* (Departamento de Ciência e Tecnologia, Secretaria da Indústria, Comercio, Ciência e Tecnologia, Governo do Estado de São Paulo, 1985).

Index

Numbers in bold refer to tables.

micropropagation of orchids 201
research strategy and planning
164–5
and restructuring of palm oil
industry 185
sugar exports: declining 211; to
USA 151
sugar industry threatened 147; by
HFCS 124
pigs, transgenic 35
plant biotechnology 12, 172
plant breeding
and increased flexibility 173
traditional ix
plant breeding techniques 178
plant oils, conversion to structured
lipids or tailored fats 151, 212
plant protection
chemicals and microbes
compared **207**
tea 102–3
plantation crops 32
plantations, large, replanting with
clonal tea 87–8
planting stocks, genetically-
engineered 156
plants, disease-free 32
plucking, labour requirements 106
policy makers, and biotechnologies
8, 9–10
politicising products 181
pollen haploid breeding techniques
190
polyclonal tea 94, 102, 103,
104, 111
employment effect 105, 108
estate 97
London auction price 110
yields below potential 96,
97, 111–12
poor people
dependent on wage employment
xi
enrichment of ix-x
harm to x
hit by biotechnology 198
threatened in the Philippines
149–53
see also rural poor

potato growers, large-scale, producing
basic seed 80
potato seed, certified, high cost of
92
potatoes 28, 55
basic seed 80, **81**
clonal multiplication of 79
in Kenya 79–87, 190, **195**;
employment potential
86–7; income generation and
distribution 82–6; labour
costs 84; need for better
marketing 92–3; output,
input and surplus 82, **83**, 84;
technological trends 79–81
micropropagation of 198–9, 218
see also seed potatoes
poultry 35
poultry feed, Nigeria 202
demand for 134
upward price trend 135–6, **136**
poultry production, Nigeria
127, 193
projected demand 133–4, **134**
projected supply **134**, 134
poverty, Philippines 148
poverty alleviation 1, 11, 85, 198–9,
206, 216
and biotechnology development
policy 11
price liberalisation 215
private sector, role of in Mexico
73–4
production costs, reduction in 200
production facilities, for micropropagation 33
profit, in BT leads to privatisation and
under-supply xii
propagation, somoclonal and
gametoclonal 172
Protein Advisory Group, United
Nations 142
protein intake, Nigeria 127, 143
protein malnutrition, combatted by
SCP 202–3
protein production and consumption, improvement
of 55
protein source, processor preferred

blending of 'low' and 'high' tech
194, 197, 218
qualified people lacking in
Mexico 71–2
requirements, increase in 120
see also scientific and technical
personnel
small farm strategies 213
developmental potential of 93
small farmers 206–7
and BT 199–200
late adopters 199–200
small-farming systems
integration of 186–7
possible marginalisation of 187
smallholder farmers, planting
polyclonal tea 111
Smallholder Tea Authority, Malawi
94, 97, 111
smallholder tea, Malawi 96–7,
110, 113
social differentiation 213
social inequality, Mexico 70, 71
social science literature, and
biotechnology 2, 3–4, 5
soft-drinks bottlers
in Mexico 125–6
using HFCS 124, 211
soya oil 167, 169, 172, 175,
178, 179
cheaper 167
soyabeans
inoculation of 43, 44–5
loss by insects, Brazil 49–50
Sri Lanka 198
stearic acid 169, 172, 174, 177
structural adjustment 216
and biotechnology innovation
10–11, 119
effect on biotechnology diffusion
215
structured lipids/tailored fats
151, 212
subsidies 215
subsistence farming 150
substitution
SCP for plant protein sources
136, 137, 138, 138
vegetable oils 187

substrates
fermentation of 174
for SCP production 59–60, 131,
132, 133, 141
sugar
Philippines industry threatened
149–51, 164
shrinking market 152
substitutions for affecting LDCs
209, 211
sugar beet, engineering of 30
sugar mills, Mexico, re-privatisation
125–6
sugarcane
Brazil 50
Mexican production, potential threat
to 124–6
use of byproducts 126
sugarcane bagasse 60, 131
microbial protein from 56
sunflower oil 178, 179
sweeteners, artificial 147, 151

Tanzania, cassava 52
tea xii
in Kenya 87–92, 195; earning
foreign exchange 93; impact
on employment 91–2, 193;
income generation and
distribution 89–91
in Malawi 94–113; analysis of
imports and exports 109–10;
analysis of production 95–8;
analysis of variance 98–102;
employment analysis 105–9;
factor proportions 102–3;
income analysis 104–5;
industry must make improvements 112
tea marketing 92
tea plants per hectare 95
tea production
factor productivity and labour
intensity 91, 91–2
future, estimates in Malawi
110, 111
labour demands 92
small-scale, not labour intensive
91